Cities of Whiteness

Antipode Book Series

General Editor: Noel Castree, Professor of Geography, University of Manchester, UK
Like its parent journal, the Antipode Book series reflects distinctive new developments in radical geography. It publishes books in a variety of formats – from reference books to works of broad explication to titles that develop and extend the scholarly research base – but the commitment is always the same: to contribute to the praxis of a new and more just society.

Published

Cities of Whiteness
Wendy S. Shaw

Neoliberalization: States, Networks, Peoples
Edited by Kim England and Kevin Ward

The Dirty Work of Neoliberalism: Cleaners in the Global Economy
Edited by Luis L. M. Aguiar and Andrew Herod

David Harvey: A Critical Reader
Edited by Noel Castree and Derek Gregory

Working the Spaces of Neoliberalism: Activism, Professionalisation and Incorporation
Edited by Nina Laurie and Liz Bondi

Threads of Labour: Garment Industry Supply Chains from the Workers' Perspective
Edited by Angela Hale and Jane Wills

Life's Work: Geographies of Social Reproduction
Edited by Katharyne Mitchell, Sallie A. Marston and Cindi Katz

Redundant Masculinities? Employment Change and White Working Class Youth
Linda McDowell

Spaces of Neoliberalism
Edited by Neil Brenner and Nik Theodore

Space, Place and the New Labour Internationalism
Edited by Peter Waterman and Jane Wills

Forthcoming

Grounding Globalization: Labour in the Age of Insecurity
Rob Lambert, Edward Webster and Andries Bezuidenhout

Decolonizing Development: Colonial Power and the Maya
Joel Wainwright

Privatization: Property and the Remaking of Nature–Society Relations
Edited by Becky Mansfield

Cities of Whiteness

Wendy S. Shaw

Blackwell
Publishing

BLACKWELL PUBLISHING
350 Main Street, Malden, MA 02148-5020, USA
9600 Garsington Road, Oxford OX4 2DQ, UK
550 Swanston Street, Carlton, Victoria 3053, Australia

First published 2007 by Blackwell Publishing Ltd

1   2007

Library of Congress Cataloging-in-Publication Data

Shaw, Wendy S.
    Cities of whiteness / Wendy S. Shaw.
       p. cm. – (Antipode book series)
    Includes bibliographical references and index.
    ISBN 978-1-4051-2912-1 (pbk. : alk. paper) – ISBN 978-1-4051-2913-8 (hardcover : alk. paper)
1. Urban geography–Australia. 2. Urbanization–Australia. 3. Whites–Race identity–Australia.
4. Racism–Australia. 5. Aboriginal Australians–Attitudes. 6. Australia–Race relations.   I. Title.
    GF801.S53 2007
    307.760994–dc22
                                                          2007018628
A catalogue record for this title is available from the British Library.

Set in 10.5/12.5pt Sabon
by SPi Publisher Sevices, Pondicherry, India
Printed and bound in Singapore
by Markono Print Media Pte Ltd

For further information on
Blackwell Publishing, visit our website:
www.blackwellpublishing.com

For my parents, Mary Shaw (1927–2006)
and Alexander George Shaw (1914–2004).

Contents

List of Figures

List of Boxes

Acknowledgements

I completed this book during a tumultuous time of professional and personal upheaval. Along with the usual thanks for bringing a book to fruition, very special thanks must go to those who supported me generally, and in writing. I am grateful to my colleagues, in particular Kevin Dunn, Jes Sammut, Chris Gibson (now at Wollongong), Ian Burnley, and the body of postgraduate geography students, who all fought to save geography (and my job) at UNSW. Thanks also to colleagues in the School of Biological, Earth and Environmental Sciences, in particular the Head of School, Paul Adam, for providing us 'human' geographers with a scholarly home. Special thanks to my friend Ry Mitchell for reading and commenting on the drafts, and for insisting that I strive to write for 'a wider audience' than the academy, and to Natascha Klocker for closely reading and editing sections of the text. Thank you to Noel Castree (series editor) and Alistair Bonnett for supporting the project, and for provided carefully considered input toward producing a more coherent text. Thanks also to the Antipode Special Series team at Blackwell, Angela Cohen, Jacqueline Scott and Rebecca DuPlessis, Stephen Erdal for the cover design and to Janey Fisher for the final copy-editing and indexing. I am also grateful to the variety of folk who gave their time and their views during interviews – the people of Redfern and Darlington.

I am indebted to friends and family members for their support over the past few years. Enriching conversations, which have provided perspective, are too many and varied to mention with accuracy but thanks to all who have discussed, listened, provided input and feedback, and argued with me. Very special thanks to Phillip Shaw who has buoyed me up throughout.

Writing and completing this book has also marked the loss of both of my parents. My father died in April 2004, and in December 2006, I suffered the loss of my mother. In a small gesture to two inspirational human beings, I dedicate this book to Alexander George and Mary Shaw.

Introduction

'This is about family!' cried the baby-strapped member of yet another newly formed resident action group. This collection of concerned citizens had grabbed the attention of other local residents, the mass media, and politicians. They had all gathered at a balloon and pram-filled public hall in inner Sydney, Australia, in June 2005. Most in the crowd were there to vent their objections to yet another proposal by the New South Wales State government to establish a healthcare facility in Sydney's Redfern area. Their problem was that the proposed facility would target what appeared to be an ever-escalating drug problem. The public outcry focused on the highly charged mix of illicit drugs (heroin mostly) and children (or babies), and a perception that the facility would simply exacerbate the problem by attracting more drug users, and therefore increase the danger to their children. The walls of the hall were plastered with large imposing posters that stated: 'No Needles Next To Children'. Newspaper clippings, also of poster size, screamed the headlines 'It's not Mr Whippy [ice-cream van], it's the Needle Van!' and 'The last thing Redfern Needs' (*Sydney Morning Herald*, June 2005).

Redfern is an inner Sydney neighbourhood that is infamous because it houses a small Aboriginal community. Eveleigh Street, Aboriginal Redfern or, as it is most commonly dubbed, **The Block**, was once just another part of the stigmatized, undesirable and run-down inner city. This inner city is now in the grip of urban renewal – of brightening, and whitening. The advent of gentrification, the refurbishment of Victorian housing stock and the redevelopment of its former industrial sites into 'apartments', has heralded a new era of revanchist mobilizations against a maligned, and heavily racialized remnant of the formerly impoverished part of the city. The consequent tensions, of a poor Indigenous settlement that suffers from a host of poverty-related social issues, set against a relatively new cohort of mostly non-Aboriginal home-buyers (and their offspring), who aspire to create a highly gentrified space in the increasingly expensive city of Sydney, seem irrevocable. The extent of the social ills of Indigenous dispossession contrasts starkly against urban renewal, and the associated desires to 'improve' urban spaces, and to ensure that they become 'safer' as well as more 'habitable'. At the aforementioned public meeting, there was no mention of who was to blame for the area's heroin problem but it is widely understood that The Block holds that dubious reputation, and has done so for the last few decades. As tensions mount over entitlement to urban space the pressure builds, and this was surely felt by the residents of the struggling Aboriginal community that day in Redfern Town Hall.

This book delves into cultural forms that have emerged with the shift away from suburban to urban living in Sydney. In many cities around the world, formerly run-down housing areas and recently abandoned industrial precincts have been rebuilt, and re-imaged as fashionable urbane chic. Like homewares, clothing and accessories, designer living spaces – be they newly built or upgraded heritage – have become increasingly popular. In Australia, a catch-cry of suburbia – where a *'man's* home is *his* castle' – is increasingly sidelined as *city* living becomes more attractive, and palatial. Accompanying such wholesale changes to inner-urban built environments, new forms of urbanism or ways of living in cities, have also emerged. These new formations have engaged strategies of exclusivity and, as I argue in the pages that follow, processes of exclusionary whiteness regardless of the city's 'multiculturalism'. Acts of defensiveness and aggression, and cultures of denial and indifference to the unpleasantness of contemporary city life are characteristic of these new urbanisms. This book is a foray into a range of racialization processes that have manifested with the re-colonization of inner Sydney.

One of the main tasks of this book is to challenge and enhance existing conceptualizations of urban change, racialization, and cosmopolitan urbanism, and I mount this challenge by arguing for a more thorough untangling of the powers of urban whiteness. *Cities of Whiteness* therefore uses its critique of responses to the presence of a particularly 'othered', and specifically racialized group – who happen also to reside at the centre of Australia's oldest and largest city – to build a case for rethinking, for de-essentializing and opening out the concept of 'whiteness'.

In the chapters that follow, I journey to the inner urban context of Sydney to demonstrate one of the main attributes of whiteness: its capacity to utilize opportunities as they arise, and to mobilize simultaneously from within a range of subject positions, as required. I use the capture of specific moments in the negotiation of urban spaces and places to unveil this capacity, particularly where negotiations occur at the expense of entitlement for the residents of The Block to live in a place designated as Aboriginal, and to carry out the tasks of daily living. The unearthing of such moments, that highlight the capacities of whiteness as they have occurred, point to new ways of untangling the processes of racialized empowerment that inhabit the evolving city.

Fields of Whiteness in Inner Sydney

Non-Aboriginal understandings about the settlement of The Block in the city of Sydney have found nourishment in the racist and racially blind historiographies of Australia. Imperialist understandings of a world of 'races', that were thought of as being either 'advanced' or 'backward', or somewhere in between, found root in the newly founded colony at Sydney Cove. From the time of contact between Indigenous peoples and the colonizers in the late eighteenth century, such conceptions have 'legitimated the exercise of power through these differences' (Jacobs 1996, 17). Although contested in many ways, understandings of 'race' in Australia (and elsewhere) still draw on these kinds of constructs. The discourses that circulate about the Aboriginal community of The Block often default to the notion of racial inferiority that arrived with the English 'First Fleet', in 1788. Continually portrayed as incapable in the modern world, Aboriginal people are often cast as uncivilized (Thomas 1994), as 'wild' and 'riotous', as unable to hold their 'grog' (that is, they get drunk easily) and around Redfern, they are often feared to be 'drug crazed' and violent.

The existence of an Aboriginal settlement at the heart of Sydney is a paradox for many non-Indigenous people. The Block does not fit the stereotype of Aboriginal Australians sitting around peacefully in nature, in the 'outback' – as is so often televized from some faraway desert place. Nor is The Block from a former time when, reiterating the words of Australia's current conservative Prime Minister John Howard, 'mistakes were made' (referring to the 'stolen generations'[1]). Its existence was formalized in the 1970s. In other words, The Block is omnipresent – it is here and now. And its high media profile, in Australia and abroad, brings the poverty and dispossession suffered by many Aboriginal people into sharp focus. In so many ways, The Block is a highly visible 'shame job'[2] for the rest of Australian society.

In *Cities of Whiteness*, I have placed a research lens on the discourses of the extended, predominantly non-Aboriginal community of the mass media, of government, and the neighbours of The Block, to tell the stories of urban whiteness. I have gathered details from various events that occurred (mostly) in the late 1990s, to build the tales of urban process that follow. Such discourses have sensationalized and demonized The Block. At the same time, these discourses are somewhat dwarfed by the extent of the redevelopment, and associated place (property) marketing, that has also occurred. Within the context of gentrification, an even more dominant but quietly embedded politics of whiteness has emerged and provided the foundation – a stage – for other performances. Acts of rampant and exclusionary enclave consciousness, by the new incumbents of gentrifying space, exemplify such performances.

Theoretical Trajectories

Cities of Whiteness has two main theoretical trajectories. First, it seeks to problematize some of the dominant thinking about 'whiteness'. The study of whiteness is now well established as a transdisciplinary scholarly field but US-based 'whiteness studies' (or 'white studies') tend to dominate, geographically and epistemologically. This book contributes to a somewhat broader-ranging engagement with whiteness that has begun to emerge within the scholarly discipline of Geography. *Cities of Whiteness* is concerned with detailing the unique intricacies of *localized* whiteness – as multiply expressed amalgams or relations of racialized empowerment – and its implications. To demonstrate this, I bring into contact the ways of whiteness and its impacts on (a group of) Indigenous peoples, who have tended to remain largely external to the ambit of current *whiteness studies* regardless of the colonial–settler context within which such studies have been, thus far, overwhelmingly generated.

A major aim of this book is to challenge the fix on 'white' ethnicity, which is so common in whiteness studies. I agree with the highly acclaimed thinker, and novelist, Toni Morrison, who has remarked on the ethnocentricity of this preoccupation, particularly the commonly declared task to 'make whiteness visible'. I have found whiteness to be much more than an ethnicity, and its association with 'Anglo' ethnicity (Bonnett 2000). Whiteness is a slippery character, a fickle entity with a capacity to expand and contract its membership, as required. I therefore provide detailed accounts of the shiftiness of whiteness, its promiscuous embrace, and its particularities that, although sometimes difficult to pin down, do give whiteness part of its strength.

The second theoretical trajectory of *Cities of Whiteness* is to identify how urban transformation and the rise of new ways of living, within these emergent landscapes, have become imbued with whiteness and its processes. Within the context of rapid urban transformation, identities, and associated urbanisms, evolve. In the city of Sydney, some of the newer ways of living have reproduced colonial formations as part of the process of re-imaging the city. For example, this book considers a range of heritage impulses driven by desires to protect certain kinds of architecture. In the case of inner Sydney, the desirable architectures are remnants of the British colonial past. But rather than simply wanting to retain these very decorative Victorian architectures, their escalated desirability has driven a whole new set of preservationist impulses that seek to restore more than pretty features. Inner Sydney is now stamped as a landscape marked with a very specific 'heritage' script that is symbolic of British settlement and its history. Preservation impulses have also emerged in newfound desires to valorize (remnants of) former industrial areas, which are swiftly transforming into Sydney's own version of that arty and exclusive Manhattan neighbourhood of SoHo, and its 'lofts'. These re-imaged inner city landscapes, with their (British) Victorian pasts and (North American) Manhattan futures, sit in stark contrast to an increasingly impoverished and out-of-place urban Aboriginal community. Subsequently, strategies have emerged that are designed to defend and consolidate the spaces and privileges of whiteness in the increasingly global, and cosmopolitan, city.

Methodological Trajectories

This book tells a story about formations of power and demonstrates, at the same time, that such formations are not bound to one colonial or ex-colonial domain. In the tradition of 'new locality studies' in Geography (see Box 1.1, page 15), that moved the humble case study to new depths, the examples provided here show how the powers of whiteness that manifest locally have wider implications. I have drawn on 'realist perspectives' that combine intensive and extensive methodologies to highlight the link between broader structures and processes in local settings (Sayer 1989, Jacobs 1993). My hope is that *Cities of Whiteness* contributes to the kind of locality study that remains alert to the broadness of issues of social inequality (such as the political economies of gentrification), while benefiting also from the richness enabled by post-structural and postmodern approaches, and an openness to the full range of methodological capacities. As I demonstrate in the following pages, the intricate strategies of whiteness that operate locally do not occur in isolation. Processes of gentrification, for example, are well understood for their capacities to marginalize and exclude along lines of difference (such as class, 'race' and gender). By turning the lens on processes that privilege, rather than focusing on exemplars of disadvantage, *Cities of Whiteness* broadens the analysis of urban change to reveal a myriad of cultural processes at work that construct and structure contemporary urban landscapes – materially and symbolically. In sum, it is the *kinds* of processes of whiteness that are detailed within these pages. These sorts of processes can be found in any contemporary urban landscape with different groups sharing, and contesting, city spaces. As Alastair Bonnett (2000, 4) identified, 'although integral to modernity…[the] form and importance [of whiteness is]…historically and geographically contingent'. Finally, and coming back to the local setting of this book, I must add that this is – in many ways – a very personal account of whiteness. Most of the events detailed here have occurred in a particular corner of the city that I call home.

A Personal Account of a Field of Whiteness

I moved into a house in Darlington just around the corner from The Block, in 1995. I had co-owned and then sold a small terrace house in another inner city area (Surry Hills) for a high price. Regardless of its poor state, the value of the Surry Hills property had benefited from Smith's 'rent gap' (Smith 1982). Of course, to owner-occupiers this was of little consequence *until* it was time to sell. The march of gentrification in our little street had suddenly gained momentum with a proposal to convert a warehouse into luxury apartments. This building backed onto our house, and the decision to convert it would change our lives.

At that time, Darlington[3] was largely a pre-gentrified neighbourhood. Having spent most of my adult life in areas that eventually gentrified, I confess to carrying a degree of pre-gentrification snobbery. I find some of the trappings of gentrification acceptable – the cafés and shops – but I have remained (somewhat hypocritically, I know) wary of the homogenizing effects of encroaching middle-classness. In Surry Hills, this had manifested in a range of resident activisms. Incoming residents ('gentrifiers') began to collectivize to drive out certain facilities and uses of urban space. These groups began to agitate against the existence of welfare services, the burning of joss sticks in an old Chinese temple,[4] 'offensive migrant house' colours (such as 'Portuguese Pink'[5] or 'Mediterranean Blue'). Darlington, I hoped, would be different. Looking back, I was (somewhat cynically) convinced that the stigmatizing presence of The Block would keep middle-classness at bay.[6] The Block's permanency (it is designated 'Aboriginal Land'), the drug and crime issues associated with the area – though no different from the rest of inner Sydney in reality – pointed to the possibility that it would retain its diversity, which included a range of age groups, ethnicities, sexualities and classes. The presence of The Block was also interesting, politically. It served to remind non-Indigenous Australia about Indigenous suffering which, in this case, was right under our noses and not out there, in the forgettable 'outback'.

Then, by the late 1990s, the unthinkable began. The presence of the Aboriginal community had stalled gentrification – as I had predicted – and buyers avoided the Darlington/Redfern area for as long as possible. Then, as gentrification cycles matured in other parts of inner Sydney (Engels 1999), and investment capital started to seek new territory, even the racialized drug and crime problems of Darlington/Redfern were not enough to deter buyers. After all, 'first stage' gentrifiers (like me) had already moved in and this vanguard included those who were amenable to the presence of The Block, or at least ambivalent to it. But this group also included those who would begin to agitate, and mobilize turf protection activities. It seemed that some were just waiting for enough 'like minded' people to move into the area; they needed a critical mass. When the media reported on an existing belief that Darlington/Redfern was about to 'take off' as a gentrification frontier,[7] some residents were ready to fight the final impediment to gentrification, and maximize their capital gain at the same time. As it happened, we 'gentrifiers' were (and are) all part of the process that legitimated Darlington/Redfern as a gentrification frontier whether we accepted the Aboriginal presence or not. In the meantime, as the area continues to change, strategies are variously deployed to 'protect' people and property from the possibility of drug and/or crime incursions.

When I moved into Darlington, the level of visible home security was striking. Be they renovated or not, houses and shops were clad in security grilles. Some houses had visible alarm systems and fences capped with barbed or razor wire. Private space was aggressively segregated from public space (Caldeira 1996). There was a general impression, and reputation, that the area was indeed a 'war zone'.

When signing the contracts to buy the Darlington house, The Block seemed to be in a poorer state than ever, and the police presence seemed to be more overt that I realized. I remember hoping that we had not made a mistake – after all, we could choose to go elsewhere, somewhere 'safer'.

Since moving into Darlington just over a decade ago, I am yet to witness any 'warfare' but I have certainly observed a series of well-planned offensives against the Aboriginal community. The history of the area includes a monumental fight against the formation of The Block (in the 1970s), and local non-Block residents continue on this trajectory. In the late 1990s, residential activism was distracted moment-arily by the advent of unwanted apartment development but as the character of the area began to change with gentrification, mobilizations focused again on The Block, which continues to be targeted to this day.

As a non-Aboriginal homeowner, and gentrifier (of sorts), I am embedded within the concerns raised in this book. The benefits to which I seem to be entitled include, amongst others, those that come with gentrification (for gentrifiers); claims to colonial heritage (I live in a Victorian house) and its protection; the capacity to join in any group (activity) that protects turf; and participate in any process that deter-mines the future of the area. Although I have associations with members of the Aboriginal community, and affiliated organizations, and have supported Aboriginal Elders in their efforts to raise concerns about the future of The Block,[8] I do not belong to that community. Rather, along with a range of other people, I generally occupy the non-Block area but I do visit when invited.[9]

As a study of whiteness, this book is not a document about the Indigenous people who live their lives on The Block. I am, however, respectfully aware of the circumstances under which their community exists. So, in *Cities of Whiteness*, my aim is to present the unearthing of specific processes of whiteness that occur and marginalize and, in direct and indirect ways, exclude the Aboriginal community from its privil-eging capacities. For the city of Sydney, a normative cultural form of whiteness, that was embedded with colonization remains, regardless of the city's multi-cultures and multicultural portrayals. It is from within such an urban setting, in an 'industrialized, western, first-world nation' such as Australia, that academic studies often 'identify, confirm, and thereby exclude, certain cultural formations as chronically marginal' (Seth et al. 1998, 9). Although several academic accounts have avoided casting The Block in this way (Anderson 1993a, b, 1998; Kohen 2000), it remains the predominant script in popular imaginings, and needs to be continually challenged. My purpose is not to silence resistances to this script from within The Block – but to unveil, with some urgency, some of the damage already wrought by activations of whiteness, and the potential for more creative activities of whiteness in inner Sydney, and beyond.

Organization of *Cities of Whiteness*

The book consists of five chapters. This introductory chapter has outlined the strands of the theoretical debates that underpin the arguments proffered throughout this book.

Chapters 2, 3 and 4 follow a thematic structure. Chapter 2 details the background to why The Block is a '(post)colonial paradox'. It outlines the context in which The Block emerged and the history of its ongoing status of 'not belonging' in the inner city of Sydney, Australia. It traces the rise of a culture of defensiveness – of urban space – and its link to a largely non-Aboriginal 'working-class' culture of besiegement in the neighbourhoods that surround the Aboriginal site. Chapter 2 also details how this status of paradox is continually reinforced by shrill discourses of the media, police and others. It foregrounds the production of specific values, that perpetuate the ongoing racialization of The Block.

Chapter 3 details one of the two key types of urban transformation that exemplify the reproduction a historical geography of racialization. This chapter documents the rebirth of inner city landscapes as desirable 'heritage' real estate, and the subsequent rise in heritage appreciation. The mobilization of whiteness in the activation of one of Sydney's 'final (gentrification) frontiers', which was stigmatized by the Aboriginal presence, is also detailed. Chapter 3 also considers the rise of heritage protectionism that works to celebrate very select histories – the 'heritage' of the colonizers that rendered the colonized territories as *Terra Nullius* (or 'empty land'). It then considers the rise of 'industrial heritage' and its role in the preservation of 'white' working-class cultural codes around the place that gave birth to The Block. Such cultural codes are being commodified and consumed by the 'new middle classes', who also invest in their protection.

Chapter 4 details more recent and somewhat subtler processes of whiteness. The production of new urban cultures that promote fantasies of escape into the past evoked through preoccupations with heritage and its protection, have become part of new imaginaries of escape more generally. New ways of living are promoted through the conversion of old industrial areas into 'New-York style' apartments (which are 'condominiums' in other parts of the world). Within these emergent urbanisms, new ways to deny or be indifferent to the presence of urban pathology have emerged. Building design and surveillance technologies have helped to minimize engagement with the local, which in turn has opened up less desirable, or stigmatized locales, such as Redfern, in the pursuit of cosmopolitan worldliness. And Chapter 5 draws together the set of threads that have rendered the city of Sydney as one of many 'Cities of Whiteness'.

Notes

1   The 'stolen generations' were the result of Aboriginal children forcibly removed from their parents and communities, who were then fostered, or adopted, into (mostly) non-Aboriginal families. Connections with kinfolk, community and cultural heritage and practices – and in many cases, Aboriginal identities – were severed. These generations of Aboriginal people had been 'stolen' from the worlds of Indigeneity.

2   Indigenous Australians (and others) use the term 'shame job' which, generally speaking, means a 'disgrace'.

3   Darlington is often included within the broader area of 'Redfern', and it is also (somewhat confusingly) conflated with the neighbourhood of Redfern.

4   Residents of Chinese origin had occupied this part of the city since the turn of the twentieth century.

5   Neighbours described our Surry Hills house as 'Portuguese Pink'. As the area gentrified, new neighbours began to pressure us to remove the paint and render – every other house was gradually stripped back to reveal the sandstock bricks, which proved to be a mistake as the bricks are extremely porous and were not designed to be without a waterproofing skin of cement render.

6   With hindsight, I now realize that I had commodified The Block in my own middle-class way, as a repellent for *other* middle-classness. Other commodifications of The Block are discussed in Chapter 4.

7   'At $508,000 Redfern may do a Paddington' (*Sydney Morning Herald*, 18 December 1996, p. 3).

8   This is a 'low profile' engagement, but I have assisted when asked. I was, by request, on the management committee of a local welfare agency that services The Block community. I provided advice for the listing of The Block on the Australian Heritage Commission's National Register (Database Number 001785, File Number: 1/12/033/0011), and provided material for a NSW state government 'Inquiry into Issues Relating to Redfern/Waterloo' (December 2004).

9   In accordance with the site's status as Aboriginal land, I do not wish to trespass so my visits to The Block occur generally by invitation – from an individual, or request to attend a public meeting, rally or performance.

1

Encountering Cities of Whiteness

Many urban settlements around the world sit on the sites of prior occupations, of peoples variously relegated to archaeological imaginations. Before colonization and the arrival of the 'white fella', the Gadigal[1] people and others of the Eora Darug clans occupied the place where the city of Sydney, Australia, now stands (Kohen 2000). While few Gadigal have survived the years of battle against the waves of invasion initiated by the English military in 1788, other Aboriginal[2] peoples were not so thoroughly dispossessed of their homelands, or places known as 'Country' around the various landmasses now known as 'Australia'. Some of the descendants of these groups have gradually re-established an Aboriginal presence in Eora country, at the heart of Australia's first and largest city. A landmark site of these recent migrations is a small settlement in the inner-city neighbourhood of 'Redfern'[3] (Figure 1.1).

In a nation that considers itself egalitarian and non-segregated,[4] the Aboriginal settlement, known as The Block, sits in stark contradiction to these national ideals. In Redfern, the boundaries of racialized segregation – the apparent 'blackness' of The Block's Aboriginality and the 'whiteness' of the mostly non-Aboriginal surrounding neighbourhood – butt against each other (see Figure 1.2, page 14). In the spaces around these borders, modalities of racialization continually evolve, and a recent expression can be found within contemporary urban transformations. In this book, I chart examples of racialization processes that have emerged with the production and consumption of two particular forms of residential transformation. One of these newly emerged housing forms is the reorientation of old housing areas into desirable 'heritage'. The other is the newest form of Sydney housing now offered with the city's recent 'Manhattanization'. This form of housing follows a

Figure 1.1   Map of Sydney's Inner City and The Block

somewhat globally recognisable trend, which is the conversion of old industrial and commercial areas into loft apartments/condominiums. With the production of these newly conceived, or re-envisioned housing forms, a range of cultural attributes that are now part of urban life has also emerged. These attributes are considered in detail within the pages of *Cities of Whiteness*.

This chapter introduces the city of Sydney, and some of the processes of whiteness that I have found to be operating within this city, which are specifically associated with these urban housing forms, and associated ways of living in the (post)modern city. As the title suggests, this book is concerned with the 'critical race study' of whiteness operating within cit*ies* – not just the one city. So the Sydney-based experiences discussed here serve to demonstrate the kinds of processes occurring in many other cities, globally. My hope is that this book will offer insights into the ways that distinct groups of people (and in the case of Sydney, non-Aboriginal people), unite and operationalize privilege and entitlement – consciously and un- or sub-consciously – through *non*-identification, or elision of the 'other'. In *Cities of Whiteness*, I discuss the ways that such unifications operate. Whiteness can be a simple form of union, or engagement, or it can be complex. Whiteness can seem permanent, but at the same time, it can be momentary, even sporadic. In this book, I unpack a variety of such unifications as they have occurred through the negotiation of urban spaces. For the 'other(s)', who in this context are the Aboriginal people of The Block in Sydney, such negotiations have occurred with recent changes in urban settlement, and the reorientation of urban areas through gentrification and redevelopment, and have provided a distinct point of capture for unveiling processes of whiteness that have hitherto remained largely unrecognized. As this book seeks to illustrate, seemingly unrelated events can be part of a wider form of metropolitan whiteness, and I will demonstrate just how mutable, particular, and self-reinforcing whiteness can be. I hope to portray that whiteness is so much more that an ethnicity, to which it is so often reduced. Rather, whiteness can work as a strategy or strategies of (urban) empowerment.

One of the tasks of this chapter is provide a background – a scholarly trajectory that has led to *this* study of whiteness. To do this I combine some of the contributions from the sub-field of **whiteness studies** (or 'white studies'), and Geography's somewhat faltering engagement with the issue. Although informed by a breadth of interdisciplinary scholarship, the study of whiteness in Geography has been somewhat sporadic. And, regardless of its transdisciplinarity,[5] the validity of whiteness as a field of research remains contested. This is not surprising given the domination of the field of whiteness studies by the United States. Additionally, 'whiteness' is all too often associated with peoples from 'Anglo' backgrounds (with some notable exceptions[6]), and one of the functions of this book is to continue the project of questioning, interrogating and opening out this limited, essentialist and, in many ways, ethnocentric categorization.

Because one of my overall aims in this book is to push the notion of whiteness beyond its usual categorizations, this chapter will provide some preliminary tastes of this goal. For me, as an Australian, one of the glaring shortcomings of the study of whiteness overall, and specifically of whiteness studies, is a general omission of Indigeneity from the orbit of concern. To begin to address this oversight I combine the process of unpacking, de-essentializing and decolonizing (the study of) whiteness, as it operates in the service of an ongoing (neo-)colonial project that further dispossesses Indigenous peoples. I do this within the context of located expressions of wider neo-colonial/neo-imperial relations of power that operate both within the city of Sydney, and in the wider Australian context. The political geographies of Sydney – like many other cities in the world – remain inseparable from the histories of colonialism, or indeed of British imperialism (cf Jacobs 1996). The Sydney-based stories presented here provide the addition of a distinctly postcolonial/imperial theoretical perspective to the study of urban whiteness.

Journeying to Inner Sydney

Contrary to popular belief, Indigenous Australians are highly urbanized.[7] This trend began, in part, with migrations to Redfern, concentrating in a tiny remnant of Gadigal country that now belongs to Aboriginal Australia. With its clutch of Victorian terraced (row) houses, The Block[8]

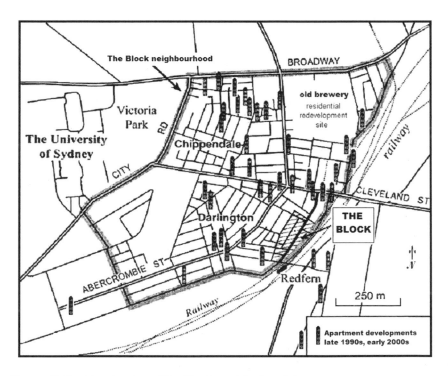

Figure 1.2 Map to show The Block and its Local Area

Box 1.1   A short history of The Block

Although not the subject of this book, it is necessary to furnish the reader with an account of the formation of the Indigenous settlement at the heart of Sydney, Australia, known simply as *The Block*. The Block straddles the suburbs of Darlington and Redfern (see Figure 1.2). It was formed in 1973 as a result of a complicated battle for Aboriginal rights. The 'Aboriginal Housing Committee' formed in the late 1960s to fight for Aboriginal housing in the context of widespread racial discrimination in the Sydney housing market and subsequent increases in Aboriginal homelessness (Anderson, 1993b, 324).

During a storm of non-Aboriginal protest the then federal Labor government, led by Gough Whitlam, granted money for the purchase of a site of 70 Victorian terrace houses for the inner city Aboriginal community. This did not occur in recognition of an Aboriginal sacred site as such, nor as a result of a land claim made under the provisions of the Native Title Act 1993, rather The Block emerged in response to a unique set of political machinations that occurred during the early 1970s (Anderson 1993a, b) when 'Redfern', a collective name for the area in which The Block sits (Darlington, Redfern and Chippendale), became the focal point of modern Aboriginal politics in Australia at that time.

The Block site consisted of mostly unoccupied or squatted terrace houses that were in a poor state of repair. A private developer bought the houses, one by one. As Anderson (1993a, 81) noted '[Aboriginal] Redfern became a sphere of Indigenous protest, an heroic site of resistance to European culture and colonialist control'. In a climate of political and trade union pressure, the owner of the site eventually sold it to the Aboriginal Housing Collective, which later became the Aboriginal Housing Company (AHC) in 1973, in order to receive the grant to purchase and administer The Block. The formation of The Block, the handing of land back to Aboriginal Australia, was a grand gesture by the new radical Labor government and it happened at the centre of Australia's 'big city' during an era of massive social upheaval. Since then, many Aboriginal services, such as the Aboriginal Medical and Legal Services, and Eora TAFE College located in the Redfern area.

Born out of struggle, The Block remains embattled. It has not been the success story of an urban village of Aboriginal self-determination that was originally envisaged. It remains a beacon for the formidable health and housing issues that exist for much of Aboriginal Australia.

(Continued)

Box 1.1 (*Continued*)

Census data from the Australian Bureau of Statistics report that Aboriginal mortality rates remain at 2.5 times that of other Australians. Asthma, Diabetes, heart problems, alcohol and other drug use are the main contributors to poor health. Life expectancy remains at 15–20 years less, with infant/prenatal mortality rates being three times the national rate, and death rates from circulatory diseases are 2.5 times greater than that of non-Aboriginal population. The 1996 census indicated that household income levels for Aboriginal people were $90 less per week than in all other households, and there was one more person in Indigenous households than the Australian average[a]. With the recent rush of renovation, restoration and re-development that has swept through the inner city of Sydney, the impoverished Aboriginal community is increasingly surrounded by affluence. Stigmatized by the presence of The Block, the gentrification of Darlington/Redfern and Chippendale did begin later than other inner city areas in Sydney but once other options were depleted, gentrification commenced in the mid-1990s. By the late 1990s, a wave of apartment development had bourgeoned. After the 'riots', in 2004, the then New South Wales State Government opposition leader, John Brogden, called for The Block to be 'bulldozed' ('Brogden's riot response: bulldoze The Block', *Sydney Morning Herald*, 16 February 2004). In the aftermath, the New South Wales Government established the Redfern–Waterloo Authority, on 26 October 2004[b*]. This authority was founded on the passing of a Bill and subsequent 'Redfern–Waterloo Authority Act 2004' and is like a government department, with its own parliamentary representative (Minister). The unveiling of its 'masterplan' for Sydney's inner city included the full redevelopment of public housing estates, and The Block. Its future is, as always, uncertain.

[a] Australian Bureau of Statistics, Population Special Article: statistics on Indigenous people of Australia, 1994 Year Book; Australia Now – A Statistical Profile, Population Special Article – A Profile of Australia's Indigenous people, Year Book 1996, Auhttp://www.abs.gov.au.

[b] 'Premier Carr Announces 10-year Redfern-Waterloo Plan', News Release, Premier of New South Wales, Australia, NSW Government, Tuesday 26 October 2004.

was handed (back) to the Indigenous peoples of Australia, in the early 1970s (see Box 1.1). This was token recognition of the dispossession of Indigenous lands, and acknowledgement of a fledgling Aboriginal politics. At the time, Redfern[9] was a virtual no-go zone for the majority of Australians, but it now sits at the heart of regeneration and redevelopment as the city of Sydney transforms from a highly suburbanized state capital into a global city (see Box 1.2).

Since its formation in 1973, the 'black capital of Australia' – as The Block was originally conceived – has coexisted uncomfortably within a predominantly non-Aboriginal urban context. As the surrounding neighbourhoods of Darlington, Redfern and Chippendale change with gentrification and redevelopment, Aboriginal and non-Aboriginal spaces are negotiated anew. This context has provided a window for studying the ways that the new occupants, who busily renovate or restore heritage housing and engage in a politics of heritage and enclave protection, variously encounter the Aboriginal presence. This book contains a group of select stories, of specific moments garnered from the past, or documented as they happened. Although such moments can stand alone, as examples of the ways of whiteness in this city, they can also be drawn together to provide a much larger picture of neo-colonial/imperial power relations. To provide a broader disciplinary context for the stories of whiteness that follow, it is to the study of gentrification and urban cultures that I now turn.

Cities as Cultural Constructions ~ Gentrification and Urbanism

Geographers have a long history of interest in processes of urban transformation. Previously studied within the domain of 'urban geography', processes of urban change such as gentrification have become part of the wider concerns of a more recent cultural politics of 'the city'. As with suburbanization, the subsequent 'return' to many inner cities with gentrification has generated dialogue, and quite heated debates. Gentrification is a term that I use with deference to the corpus of work devoted to the concept. I remain, however, somewhat ambivalent about its conceptualization and believe, as with the term 'whiteness', it needs to be used with care. 'Gentrification' is due for a conceptual overhaul but to do this properly would require writing a different book.

Box 1.2   Sydney settlement in the twentieth century

The majority of Australians live in suburbs. Mass migration to tracts of land that became 'suburbia' occurred during the housing shortages after World War II, which loosely resembled the North American phenomenon of 'white flight'. Inner-city slum clearance, government and developer initiatives, improved transport links and the re-domestication of women – after war duties had taken them out of their homes – were all part of the building of Sydney's suburbs (Murphy and Watson 1997). Historian, Shirley Fitzgerald (1987, 42), observed that 'literature advertising the garden suburbs appealed directly to class snobbery, [and] depreciat[ed] the older [inner-city] suburbs'. As Sydney expanded west in the 1950s and 1960s, ownership of a house on a plot of land away from the tenements of the industrial, dirty, inner city, dramatically changed the face of Australian domesticity (Murphy and Watson, 1997).

After World War II, another settlement event occurred. The first generation of post-war migrants were quietly moving into pockets of the otherwise largely undesirable parts of Australian cities. These inner-city pockets were referred to as 'ethnic ghettos' (Burnley and Murphy, 1994) and government policy makers assumed that concentrations of ethnic groups signalled a bourgeoning underclass. As tenancy rates soared in the inner cities, reaching 90 per cent (prior to gentrification) in a nation with an increasingly suburbanized home-ownership ideal, the suburbs represented affordable respectability for the majority of Australians (Fitzgerald 1987). Thanks to the White Australia Policy of 1901, Australians at that time were overwhelmingly of British ancestry.[a] Consecutive post-war governments provided home ownership initiatives that assisted suburbanization, and for those leaving the inner city behind, home ownership in the suburbs promised upward mobility. Many became the new 'middle classes', and in the 1960s and 1970s these groups began to migrate (back) to inner Sydney, with its romanticized Victorian landscapes of heritage houses.

[a] Castles, Kalantzis, Cope and Morrissey (1988, 19) reported that by 1947 'Australia had the lowest proportion of overseas born ever recorded for the non-Aboriginal population'. At the time, Aboriginal people were mostly relegated to non-urban areas.

A few words on 'gentrification'

Rather than rehearse the well-documented history of gentrification scholarship (see instead Lees 2000, Atkinson and Bridge 2005), I will move beyond a theoretical impasse that occurred in the 1980s, when gentrification debates became polarized between economic and cultural imperatives. By the 1990s, the cultural politics of gentrification, or as Loretta Lees (1999, 127) put it, the usefulness of the 'tensions between different theoretical positions', had gained recognition.

In my view, the great theoretical leap forward in gentrification thinking occurred with the inclusion of localized cultural politics (Lees 1994, Jackson 1995, Jacobs 1996, Redfern 1997), particularly when considered within the context of broader, more global, political economies (Smith 1996). Critiques of the larger economics of gentrification led to the understanding that gentrification was not a homogeneous process; events were not necessarily replicable from city to city, or from country to country (cf Engels 1999). For example, it was realized that displacement of the 'working classes' was not a *necessary* outcome of gentrification, regardless of the implication of the term. Additionally, a predominantly 'cultural' theoretical perspective that did not restrict its understandings of gentrification to analyses of power relations along class, economic or other lines, provided some other useful insights. David Ley (1974, 1980, 2004) and Caroline Mills (1988, 1993) provided careful analyses of consumption in gentrifying Canadian cityscapes that helped to challenge existing class-based assumptions about urban transformation. With hindsight, they also demonstrated another very useful point. Their studies demonstrated the problem of privileging one explanation for gentrification, and in these studies, a cultural analysis held singular priority. Mills' studies of 'packaged' cultural identities, as marketed by property developers in Vancouver, provided detailed accounts of the development and marketing of an inner city. A social process was identified, but was limited to drawing out a relationship between developer (producer) and purchaser (consumer) of culturally packaged real estate. Although a groundbreaking analysis of the motivations of developers and (to an extent) buyers, the impacts of such negotiations on other groups of people, or services, were not included in the analyses, whether they existed or not.

Lees (1994) suggested that analyses of capital and culture do need to be complementary in order to venture beyond the culture/class divide, and more inclusively account for gentrification processes. Jackson

(1995) noticed that culture (in a broader than conventional definition, which tends to refer to 'high culture', to theatre and art galleries) was inseparable from economy. Using two American urban examples as case studies, he identified the 'cultural encoding' within the economics of investment in the built environment. Elsewhere, Engels (1999) identified the need for more cross-cultural research to appreciate the complexity of gentrification, adding that gentrification needed to be observed, over time. 'Gentrification' is not a unified term. It may refer to very specific localized processes or to broader urban changes. More recently, considerations of globalizing urban cultures (Eade 1997) have tied global economics to consumer tastes (Bourdieu 1984, Collins 1995), and the production of place-specific identity traits have also been increasingly linked to processes of capital. As cities, and city life, are various reshaped by corporate capital, developer-driven urban transformations have become part of the broader study of the city (see Thrift and Walling, 2000, for a review).

So in this book I consider 'gentrification' in a very broad sense, as an outcome and indicator of economic shifts that have occurred at the level of the local, the city and beyond (see Box 1.3). But equally, gentrification is a process that is culturally encoded *by*, and part of the process *of* (re-)shaping the residential city. *Cities of Whiteness* details the (re)population of the older inner city by former suburban dwellers, and by newcomers, who 'revitalize' housing stock. Processes of urban renewal that have created new residential landscapes are also part of the overall gentrification of inner Sydney.

To conclude this short section on gentrification, I must stress that in the world outside the academy, the term is now well entrenched in the

Box 1.3   New locality studies

A reinvented case study approach has accompanied the cultural turn in Human Geography, where the use of a case study is not restricted to the location. This approach assumes that it is the processes in a case study area that are significant, rather than the area as a case study, in itself. This kind of locality-specific case study has numerous origins in recent geographic thought. The positivist era had relegated studies of location to structuralist studies of, for example, capitalist production (Jackson 1991a). The 'new locality study' approach promised a better understanding of processes, such as global capitalism, through the observation of variety in globalization's localized consequences (Duncan 1980, Massey and Allen 1984). Because localities differ in the way they respond to, resist, and constitute

Box 1.3   (Continued)

economic restructuring, 'new locality studies' in Geography enabled new understandings of social relationships and structures.

To find ways that local cultures and (to continue the thread) global economics might interact or constitute each other, geographers also looked beyond their discipline. As a social science that had spent a long time considering the materialities of economics, new cultural geographies needed to grapple with the cultural particularities of place and to 'identify the qualities that structure that experience [of place]' (Jackson 1991a, 222).

The embrace of the cultural turn in Geography has also re-invigorated qualitative methodologies in its search for cultural data. The contribution of Clifford Geertz's 'thick descriptions', the use of local knowledge to deepen and ground narratives of interpretation, and experimental methods of ethnographic description, used to address issues of 'representation', are well recognized in the social sciences (Jackson 1991b). According to Peter Jackson (1991a, 215, 219), the new phase in locality studies in Geography combined:

> Alternative theorizations of 'local culture' [which] draw...on concepts of cultural politics (from Stuart Hall), structures of feeling (Ramond Williams), cultural capital (Pierre Bourdieu) and local knowledge (Clifford Geertz)...where 'cultural politics' [is] defined as the domain in which meanings are constructed and negotiated...the cultural is always...political.

The new locality studies have not only remained alert to a range of political issues that were brought to the geographical research agenda through structuralism, they have drawn on the contributions of poststructuralism and postmodernism in Human Geography and other disciplines such as sociology, anthropology, postcolonial and cultural studies.

When Andrew Sayer (1989, 256) noted the swing back to the use of 'case studies' in human geography, he identified it as part of a more general 'empirical turn' and, more importantly, a growing concern for what 'real people', as opposed to 'the ciphers of social theory', were doing and thinking. Building upon anthropological case study approaches – which tended to search for generalities (Mitchell 1983) – the more recent geographical studies had became far more context specific. Such approaches were 'distinguished by...self-conscious engagement with social theory...[and drew] upon different, mainly

(Continued)

Box 1.3   (Continued)

realist, rationales' (Sayer 1989, 253). According to Jane M. Jacobs (1993, 830), such realist perspectives provided the link between broader structures and local processes in local settings, and 'a commitment to a range of qualitative approaches'.

By 2005, this kind of approach had become common within Geography, and in that year Noel Castree called upon geographers to *really* think about this approach, and its plausibility. He reiterated one of Andrew Sayer's concerns, about the divide between the idiographic (particular cases) and nomothetic (general laws). According to Sayer (1989, 255) the fundamental question remained 'how far, or at what depth, are social structures and processes context-dependent?'

For the divide between the idiographic, and nomothetic to be properly collapsed, Sayer suggested that regional geography/locality studies be theoretically informed *and* empirically based. According to Sayer (1989, 257), ethnography has 'pulled towards contextualizing explanations or interpretations', whereas political economy tended to use analytical explanations that are 'independent of context...[and therefore] law-like'. The *new* locality studies offered 'methodological challenge[s]' because they would attempt to 'hold the two [political economy and contextual explanations] in tension' (Sayer 1989, 257).

More general criticisms have come with critiques of the cultural turn, more generally. For instance, Blair Badcock (1996, 91), has expressed concerns about the ' '[l]ooking-glass' views of the city'. Sayer (1993) remarked that some researchers were 'more struck by the chasm between "the esoteria of postmodernism" [than] "what is happening outside" academia to ordinary people in our communities'; and Vera Chouinard was concerned that 'working class and other disadvantaged groups...[were] often curiously absent from the landscapes...of "consumption", "spectacle" or "power"... represented in postmodern cultural geographies of the city' (cited in Badcock 1996, 92). Badcock continued with a point about gentrification, arguing that the study of gentrification had become a 'fixation...at the expense of countless unexamined neighbourhoods where the impact of structural decline, disinvestment and the withdrawal of social services is equally profound' (Badcock 1996, 92). Furthermore, he warned that 'concern about the level at which the imagery and façadism registers in the consciousness of the imbibers...[means that] far too much can be...read into the meaning of

Box 1.3 (*Continued*)

postmodern landscapes given their potential for duplicity' (Badcock 1996, 94).

In *Cities of Whiteness,* I have attempted to demonstrate that (although the book is not about gentrification *per se*) this kind of research – on consumption of housing and lifestyle choices in the postmodern landscape of inner Sydney – does genuinely address issues of marginalization. I have tried to broaden the usual gentrification/urban transformation focus to show the very real links between such processes *and* their marginalizing effects. It must be said that the ongoing marginality of the Aboriginal community in inner Sydney is not external to the variety of cultural processes associated with gentrification. Quite the contrary, as I hope I have demonstrated, such processes are operationalized to the benefit of some, often at the expense of other(s). The Aboriginal 'others' are variously present and absent in the imaginaries of lifestyle choices as they are in the material effects of urban change. Understanding transformations of places, including gentrification processes in cities, is vital to understanding complex processes of racialization and consequent marginalization. The constructions of imagery, and façadism, as documented in this book have revealed both direct *and* duplicitous efforts that have variously built upon existing racialized power relations. I argue that through image-making of new urban lifestyles, for instance, new sites of consumption have emerged that serve to reinforce existing neo-colonial processes in the privileging of whiteness.

According to Ed Soja (1999, 65), the 'cultural turn' in turn 'culturalized' political economy. As such, there was not a sudden shift in ideology, nor a wholesale leap from modernism and structuralism to de-structured or 'flip' postmodernism (as Sayer predicted in 1993). Indeed, the use of the term **postmodernism** has tended to be over-zealously embraced. Rather than continue a somewhat circular debate, Driver (1995, 129) has suggested that there is 'much more to be gained from serious engagements with the specific philosophies, politics and methods too often hastily subsumed under that heading [of postmodernism]'. In short, geographers are quite cognisant of the usefulness of methodologies that include the reading of meaning in the salvaging of old factory façades. Remaining vigilant to the possibilities of method, and being inclusive of the full range on offer, transdisciplinary approaches remain open to the inclusion of the specifics of local encounters *within* the context of broader politics.

marketing parlance of real-estate agents. As Goodwin (1993, 147) remarked, '[t]he selling of an urban lifestyle has become part and parcel of an increasingly sophisticated commodification of everyday life'. With the emergence of new urban lifestyles that are part of the cultural capital of gentrifying, or re-imaged and rebuilt areas (such as London's Docklands, Manhattan's Battery Park, Melbourne's Docklands and Sydney's Pyrmont), urban research, and its theorization, now includes the lifestyle concept of 'urbanism'. So, in the following, I contemplate this recent (re)turn to thinking about urbanism, or urban social life, particularly within Geography.

Urbanism and urban lifestyles

> Urbanism is about living in cities; it represents a way of life affecting the majority of people . . . (Forrest and Burnley 1985, 1)

The term **urbanism** has re-emerged in geographical (and other) considerations of the city. Largely relegated to the pre-positivist era, and locked away in the writings and musings of Simmel (1995 [1903]), Benjamin (1978 [1935]) and Wirth (1995 [1938]), urbanism – *as a way of life* – was largely overlooked during the era of positivism[10] in Human Geography (however, see Harvey 1972). As Mike Davis (1990, 2) observed, '[t]he ways people lived in cities merely reflected the social organisation of a particular economic order: capitalist urbanism was . . . fundamentally different to socialist urbanism'. With the rise of 'new' urban sociology (Milicevic 2001 for review), post-structuralist methodologies and the cultural turn of the 1980s and 1990s, geographers (and others) began to think about urbanism and urbanity afresh. The renewed interest in urbanism is also evident in the discourses of urban/ town planning, and architecture, with the rise of New Urbanism. But this 'outcome-based view of planning based on a vision of a compact, heterogeneous city' (Fainstein 2000, 451) is prescriptive, and quite distinct from geographers' considerations of urbanism(s) exemplified within the recent surge of geographical interest in formations of urban lifestyles.

The (re)turn to urbanism as a valid terrain for scholarly contemplations, is not without its critics. The incorporation of cultural factors into analyses of cities has raised suspicions about the political implications of de-prioritizing economic (and class) factors. For example, when Michael Dear and Stephen Flusty's (1998) 'Postmodern Urbanism' was published in the *Annals of the Association of American Geographers*, a controversy in social theory was ignited. In response to the piece, Robert Lake (1999) warned against tying urbanism to postmodernism, and a heated debate began between self-professed postmodernists (Dear and Flusty) and others (who defy being labelled as 'modernists', as such). Beauregard (1999), Lake (1999), Sui (1999) and Jackson (1999a) launched separate, but equally blistering critiques of Dear and Flusty's complicated but not particularly postmodern version of 'Postmodern Urbanism', published in *Urban Geography*. Dear and Flusty (1998, 50) had claimed that a Los Angeles model of urbanism was 'distinguished by a centreless urban form, termed "keno capitalism" '. This reversal of the power of capital, where the 'hinterland organizes the centre', was the basis to claim that Los Angeles urbanism had provided a 'radical break' in understandings of the ways that cities develop (Dear and Flusty 1998, 50). This conceptual fracturing of what Dear and Flusty described as the Chicago School's 'classical modernist vision' of the industrial metropolis, created an intellectual tug-of-war with the retort that such claims contained theoretical 'dead ends' and 'ethnographic voids'. Urbanism became the object over which yet another postmodern-geographies versus other-human-geographies debate could rage. But a side benefit to the quarrels[11] was that they highlighted the need for geographers to re-engage with the notion of urbanism – 'postmodern' or otherwise. Far more dispassionately, Clark (2000, 17) provided this definition:

> Urbanism is the name which is most commonly used to describe the social and behavioural characteristics of urban living which are being extended across society as a whole as people adopt urban values, expectations and lifestyles.

In the same work (page 19) Clark urged that urbanism be considered along with the study of urban growth and urbanization. He noted that 'the spread of urbanism is linked to the emergence of a global society made possible by developments in telecommunications and mass media' (Clark 2000, 15), and that globalizing cultural indicators, such as commodity brandings, were having an increasing influence on urbanism.

> Social values and relationships have ... become similar across the world. They have lost their connection with a specific place and are constructed and spread by the mass media ... the spread of urbanism is linked ... to the emergence of a global society (Clark 2000, 15, 21).

This highly generic version of urbanism – particularly given the variety of ways that city spaces are occupied and used – does highlight the issue of the mobilization of power within contemporary urbanism(s). Even the protagonist of 'postmodern urbanism', Michael Dear (2000, 1) commented that:

> The creation of different kinds of urbanism, characterized by edge cities, gated communities, and a global hierarchy of new 'world cities' . . . is a key to understanding the burgeoning geopolitical order.

Although not specifically defined, the (very un-postmodern) concept of a 'bourgeoning geopolitical order' ignited contemplation of urban lifestyles that rely on the concept of a 'global society'. Pushing this idea a little further, *Cities of Whiteness* considers how the consumption and protection of colonially-referenced architecture, embedded within Sydney's urban fabric, has influenced the formation of contemporary, globally-referenced, urban identities. In addition, New York style 'loft living' has added another dimension to the new kinds of urbanisms found in Sydney. In the chapters that follow, I unpack some of the specificities and consequences of what *appear* to be globalized (generic) expressions of urbanism in Sydney. In so doing, and in line with its title, *Cities of Whiteness* also interrogates Alastair Bonnett's (2000, 3) conception of global(izing) white identity formations.

> White identities are, if nothing else, global phenomena, with global impacts. Indeed, the nature and implications of their local manifestations only come into view when they are understood as local . . . [I am advocating] an attempt to engage the international and comparative diversity of whiteness.

Rather than analysing local expressions of a globalizing white urbanism, however, I engage instead with Bonnett's idea of 'comparative diversity', which has helped to unveil fantasies about *imagined* cosmopolitan urbanisms. Such fantasies have been conjured within local understandings in Sydney, about far-flung places such as Britain (through understandings of heritage) and New York (through New York style 'loft' development), past and present. The convergences between the spread of contemporary urbanisms to locations such as Sydney (the notion of 'spread' itself revealing another fantasy of a perpetual 'centre' and 'periphery' of some civilized world) and consequent identity formations that perform their own processes of consolidating whiteness, have become highly visible in juxtaposition to the existence of an impoverished

Indigenous community. Existing at the heart of inner Sydney, The Block has been excluded in many ways from the fantasies and realities of cosmopolitanism. Its ongoing struggle is to simply exist. This Australian context has provided a unique window for observing the production of *particular* versions of new (white) urbanism(s). To make sense of the concept that underscores every aspect of this book, and to give context to my overall aims of de-essentializing and de-ethnicizing it, I now move to the idea of whiteness – its history, and current theoretical trajectories.

The Birth of Whiteness Scholarship

For the purposes of this book, I have drawn on a history of the study of whiteness from both within and outside of the scholarly discipline of Geography. My first task in this tracing of whiteness thinking and research is to précis its emergence as a research field, and to consider the various ways in which it has been conceptualized.

Largely based in the US, critical studies of whiteness appeared in the late 1980s (Dyer 1988) in a range of disciplines such as literary criticism and cultural studies. By 2000, a bourgeoning quasi-discipline was dubbed *whiteness studies* (or 'white studies'), with a raft of undergraduate courseware and publications on issues associated with post-slavery segregation and racism, white supremacy, and Hispanic migration (Allen 1994, Dyer 1988, Frankenberg 1993, Ignatiev 1995). But, rather than simply mapping (racial/ethnic) difference,[12] more nuanced studies of racialization (and therefore whiteness) have explored the processes by which difference is constituted (cf Fincher and Jacobs 1998). Building on this idea, some studies of whiteness have identified the processes of empowerment that have enabled 'white' ethnicity to designate difference, that is, to allocate other ethnicities. However, understandings of the character of whiteness do tend to be based on an assumption about 'Anglo' ethnicity,[13] and this needs further examination. Where critiques of the Anglocentricity of whiteness have emerged – such as from within the discipline of Geography – they have remained largely on the periphery of the juggernaut of whiteness studies.[14]

'Critical race studies' and the rise of 'whiteness'

There is a strong tradition of interest in issues of racial segregation and racism within Geography (for a review see Anderson 1998). In North

America, early studies tended to concentrate on the formations of 'ghettos', with pioneers in the study of racial segregation including DuBois (1899) and Weaver (1948). Much later, Ceri Peach (1993) identified the three main approaches to segregation research as Morrill's 'positivist' project, Harvey's 'Marxist' project,[15] and Ley's 'humanist' project.[16] These three studies were the hallmarks in Geography's 'critical race studies'.

In Peter Jackson's *Progress in Human Geography* report (1985a), the state of critical race scholarship was summarized within four categories. 'Racial and ethnic segregation' research (after DuBois and Weaver), was concerned with the patterns of urban segregation and ethnic concentration. The second category was concerned with 'race and ideology' and concentrated on 'structured inequalities of power... [such as] the relationship between 'blacks' and the police (Jackson 1985a, 101). The third category of research, on 'riots and rebellion', consisted of responses to moments of political upheaval associated with racialized groups (for example, the 'Brixton Riots'). By the mid 1980s, critical race research was increasingly concerned with the 'nature of ethnic politics' and interrogation of, for instance, the notion of the 'ethnic problem' (Jackson 1985a). At that time, Jackson also observed a gap in research relating to the analysis of 'white society'. Noting that 'white institutions... generat[e] racial inequality' (Karn in Jackson 1985a, 100), Jackson called for increased attention to be paid to the (white dominated) mass media (Gabriel 1998) and flagged the need for geographical research to problematize whiteness.

Somewhat ironically, little published academic work existed on the theme of whiteness in the mid-1980s yet, as bell hooks (1992) had observed, in the US, whiteness has been under critical observation since the times of North American slavery. According to hooks (1992, 338), 'details, facts, observations, psychoanalytic readings of the white "Other" ' had been a survival strategy of African Americans since their arrival on North American shores.

In another *Progress in Human Geography* report, Helga Leitner (1992) reported that although there had been a shift from considerations of 'race' to more nuanced geographies of racialization, she remained 'disheartened' by an ongoing disciplinary proclivity for reducing 'the notoriously complex and elusive concept of cultural assimilation... to a single variable [race]' (Leitner 1992, 106). Geography may have taken a useful turn in theorization but in practice, simplistic racial categorizations prevailed. Leitner did note the emergence of a second body of research that used qualitative analyses from case studies and broader conceptions of culture that included considerations of class and gender (S.Smith 1989, 1990). Most significantly, these studies sought to contextualize racialization.

By 1998, Kay Anderson had identified a worrying trend in racialization research. She remarked on the persistence of 'an ordered (racialized) reality whose subject positionings [remained]...fixed and undifferentiated' (Anderson 1998, 206). Consequently, she argued, these 'neat stories of unilateral hegemony' (Anderson 1998, 210), were downplaying and homogenizing difference. Meanwhile, some geographers were attempting to encompass the concept of difference by reorientating scholarly attention to the project of identifying the *politics* of difference (Dunn 1993, Fincher and Jacobs 1998, Young 1990).

Difference, in Australia, focuses mainly on migrations. Since colonization in 1788, migrants have come in waves. First came the British, which may seem an obvious point but is often forgotten. After this colonizing wave – which became the point against which difference in Australia would thereafter be identified – 'others' began to arrive. Migrations from China began during the gold rushes of the 1800s. The next main wave of migration was from Southern Europe after World War II, and then from South-East Asia after the war in Vietnam/Viet Nam, in the 1970s. The most recent migrations have been from 'the Middle East'. With the English as the arbiters of difference, subsequent arrivals have provided pivotal moments for constructions of difference in Australia. Indigeneity, however, has remained glaringly absent from the discourses of difference and policies of multiculturalism for several reasons (Curthoys 2000, Stephenson 2003). The Aboriginal peoples of Australia are – quite rightly – perceived as different to and separate from the multicultural mosaic of identities in Australia, and are themselves particularly sensitive to being categorized as part of this mix. 'Multiculturalism' invariably refers to the cultural distinctiveness of 'migrants'. A consequence of this separation, however, is that studies of Indigeneity have too long remained on the fringes of Anthropology and Archaeology, as a somewhat cross-disciplinary form of 'Aboriginal Studies'.[17] One landmark study did shift some of the emphasis from studying outback Indigeneity and 'traditions', to urban settlement. Fay Gale's *Urban Aborigines* (1972) gave rise to studies about Redfern in Sydney (Anderson 1993, Shaw 2000) and Perth in Western Australia (Fielder 1991, Jacobs 1996, Jones 1997, Mickler 1998). More recent concerns about 'difference' have shifted along political (and therefore research funding) lines to encompass the (apparently) more pressing issues of 'border security' and the status of asylum-seekers.

As the embrace of 'difference' has expanded, one lens of research has moved towards the study of majority groups and the dominant forces that have prescribed the norms or benchmarks against which to identify difference. This has led to identification of 'whiteness'. In his ground-breaking explorations, Ghassan Hage (1993) has argued that both the policies of multiculturalism and the recent acknowledgment of Native Title rights[18] have officially legitimized and sanctioned the presence of minority groups, in Australia. This recognition, that minority groups have certain entitlements, has in turn incited a siege mentality within Australia's 'dominant white culture' (Hage 1993). Extrapolating from these observations it is not difficult to identify a similar sense of 'siege' around inner Sydney. This sense of siege is felt at the local through to the national and even international levels, and was exemplified well in reports of the 'Sickening ['race'] Riots' that 'Rocked Sydney' (*Sydney Morning Herald*, 16 February 2004, 1). These so-called 'race riots' were a response to the death of a 17-year-old Aboriginal boy who was killed when his bicycle crashed into a fence while he fled from police.[19] Such events, and the exaggerated reportage that followed the 'Redfern Riots',[20] are not new to inner Sydney, or Sydney more broadly as the recent incursions (more 'race riots') in beach suburbs, attest.[21] In the case of Redfern, small outbreaks of rage and frustration do occur, and are often a response to the status of Indigeneity in Australia, magnified through overt poverty and associated social ills. It is from within this broader neo/post-colonial context that I have observed localized render-ings of racialized difference that are so intrinsic to my understandings of the workings of whiteness.

From whiteness studies to geographical studies of whiteness

As whiteness studies gained recognition and status in US, Alastair Bonnett had begun to carve out spaces for the study of whiteness in Geography, over in the UK. Bonnett's evolving discourse on whiteness (from 1992 onward) provided a framework for making sense of the variety in racialization, and marginalization processes in Geography. Bonnett offered a way to pursue the interrogation of whiteness with his pioneering study of anti-racist strategies in school education (Bonnett 1992, 1993). By 1997, however, he detected a deeply rooted disciplinary stasis that he believed was not entirely accidental. Bonnett may have persistently steered geographers towards the idea of whiteness, but the innovative scholarship on the construction of dominant categories and majority groups, remains largely outside of the discipline of Geography. As Bonnett (1997, 193) identified, in 1997, 'the racialised subjects of geographical enquiry have remained . . . the same, namely the activities and inclinations of marginalized ethnic groups, most especially non-Whites'. This 'effacement of the "white" subject', and the continued focus on constructions of the 'other' (see also Robinson 1994) has reflected more than a disciplinary unwillingness to engage. After all, geographers are well rehearsed in identifying political positionalities (Jackson 1991b). According to Pulido (2002, 46), Geography was still 'informed by experiences of whiteness',[22] in 2002.

To make sense of the breadth of studies of whiteness, and to position the potential for its study within Geography, Bonnett (2000) offered four broad categories. These were the studies of 'anti-racism in white areas', the literatures of 'white confession', the engagements with 'excluded whites' and the studies of the 'historical geographies of whiteness'. Within this last category, analyses of the social contingency of whiteness, and critiques of the category 'white', formed an emergent research 'school', with two broad approaches. One included the now famous works of Allen (1994), Ignatiev (1995) and Roediger (1991), who based their analyses of whiteness on class, and the second, and potentially most useful approach for Geography, 'stresse[d] the plural constitution and multiply lived experiences of whiteness' (Bonnett 2000, 121).

Regardless of its potential, however, research on whiteness has tended to suffer from guilt by discursive association with the bulk of whiteness/ white studies. In a progress report on the state of 'anti-racist geographies', Catherine Nash (2003, 641) reiterated a common concern about the 'image of monolithic whiteness' and mentioned one 'useful' study of 'Not In My Back Yard' (NIMBY) conflicts (Wilton 2001). The authors of one paper, on an 'epistemology' of whiteness (Dwyer and Jones 2000), faced the accusation that they had simply reinforced the 'trans-historical, essential, asocial and universal character of unmarked whiteness' (Nash 2003, 640).

So regardless of Pulido's (2002, 45) observation that studying whiteness 'may be . . . a less problematic area of inquiry' than studying the 'other' within such an ethnically 'white' discipline as Geography, there is room for a more thorough geographical critique. I can only assume that the disciplinary preference *not* to study whiteness is due to the perpetuation of an understanding about the perceived nature of whiteness – particularly its ethnic rendering. In other words, I am suggesting that the awareness of the problems associated with much of the research encompassed by whiteness studies has somewhat stigmatized the concept, and stymied engagement with its potentially rich fields, within geographical imaginations.

Problematic whiteness

A major concern about 'whiteness' remains its ongoing affiliation with a powerful hegemonic ethnicity. Robyn Wiegman (1999) observed that within research, the term 'whiteness' overwhelmingly referred to a racial or ethnic category. It is therefore prone to the production of meta-narratives about universal modalities of dominance. Another, almost contradictory worry was that explanations or illuminations of racial division and domination, might (inadvertently) reinforce whiteness. Pulido (2002), for instance, has reiterated a common concern that there is a risk of glorifying whiteness, and studying it at the expense of research on constructions of otherness (i.e. other than whiteness). Another question has been raised about the intersection of whiteness with other axes of difference, such as class and/or gender, in the production of social and economic inequality (summarized in Oliver 2002, 1272).

One of the main limitations of US-based studies of race and racialization, and therefore whiteness, has been identified as the reliance on a specific point in North American history, which has dominated understandings of segregation. According to James O. Horton, the invention of 'race' in the US has been intrinsically linked to the time of the Declaration of Independence, and Thomas Jefferson's 'Notes on the State of Virginia', which stated:

> I advance it, as a suspicion only, that the blacks, whether originally a distinct race, or made distinct by time and circumstances, are inferior to the whites in the endowments both of body and mind.[23]

Although a monumental point in US history, this preoccupation has detracted from more complex histories of colonialism(s) and imperialism(s) that have survived into the present. Such histories have created current patterns of segregation, and held neo-colonial formations in place.

Wiegman (1999, 19) also identified that a 'repeated appeal to the minoritized, injured "nature" of whiteness' (cf Brown 1995) appeared to be in step with the existence of the largely North American fields of 'Ethnic [and Black] Studies', which are commonly associated with university departments. One 'school' of whiteness studies has focused on the 'marginal' status of those identified as 'white trash' (Wray and Newitz 1997). This 'school' has laudably included class analyses within studies of 'white' ethnicities (Winders 2003), and this is indeed a field in need of sympathetic inquiry. However, Wiegman (1999) has warned against any tendency to sympathise with the 'minoritarian positionings' of, for instance, 'white' racial supremacists. Obviously, the violences associated with this brand of 'difference' (Back 2002, Ezekiel 1995, Fredrickson 1981, Kimmel 2000) have themselves invoked the exclusions experienced by the perpetrators. Of concern is the potential to appropriate the claim of marginality due to 'white' ethnicity. This is a disturbing converse politics to the study, and exposure, of the marginalizing, and at times overtly violent, powers of whiteness.

For me, remaining vigilant to the aforementioned problems with studying whiteness is part of its demand for attention, and therefore my engagement with its formation. I have found 'whiteness' to be an extremely useful heuristic tool, regardless of its discursive associations with an essentialized, hegemonic ethnicity or potentially monolithic 'power framework' (cf Winders 2003). As I seek to demonstrate in this book, when uncoupled from such associations (the study of) whiteness can provide the capacity to encapsulate sets of context-specific processes and performances of, at times, very subtle forms of racialization. From my research experiences, I have found whiteness to involve processes that privilege and/or dominate. In a city like Sydney, such processes often do benefit a group that is identified as 'Anglo' (whatever that may be), but not solely. As I will demonstrate, whiteness is mutable and flexible, and so are the boundaries of its memberships. The important point to remember is that the link between the ascribed ethnicity, and the processes of empowerment, constitute a handy marriage in the pursuit of the latter.

In the remainder of this chapter, I will use some of Toni Morrison's writings to explore the idea of subject positionings that both privilege *and* de-privilege the designation of ethnicity to 'whiteness'. I then trace some of my early encounters that raised questions about my own assumptions about 'white' ethnicity. Additionally, I will introduce some of the ways that Indigenous peoples, in Australia, have been subjected to the incursions, and power, of whiteness. But rather than 'studying' Indigenous peoples, I do this by following the path of studying whiteness, observing non-Indigenous educational structures, and uses of space, with the addition of a 'postcolonial perspective' (see Box 1.4). This perspective has helped to provide an understanding of the post-/ neo-colonial contexts within which Indigenous struggles occur. It has also provided an example of a strategy for moving the study of whiteness out of its usual domains.

Box 1.4   Through a postcolonial lens...

The field of postcolonial studies, in Australia (Jacobs 1996, Gelder and Jacobs 1998, Gandhi 1998), and elsewhere (Said 1978, Spivak 1985, 1987, 1990 and Bhabha 1990, 1994, 1998), has shaped the theoretical, and political, positioning of this book. Although critical race studies and postcolonialism have tended to remain somewhat separated in Geography (Jackson and Jacobs 1996), more recent studies have paid attention to the importance of colonial histories to racialization processes in the present (for example, Jacobs 1996). Jackson and Jacobs (1996, 3) identified that 'postcolonial studies have helped us attend to the complex ways that the past inheres in the present', and this marriage between postcolonial and racialisation research has enabled me to acknowledge the role and uniqueness of colonial histories and their influences on race relations in the present, in Australia.

On the ground, in the neighbourhoods surrounding The Block in Sydney, a colonial past is inscribed in the present in very specific ways. Non-Aboriginal people continually encounter the Aboriginal presence that colonization sought to dispossess and consequent urban settlement worked to displace. In subtle ways, such encounters invoke the colonial past, calling it into the present and requiring it, and its consequences, to be negotiated anew. Indigenous people were banished from the original settlement of Sydney (Reynolds 1996), and the formalised 'return' of the Aboriginal presence to the city, in the early 1970s, which took the form of a specifically designated Aboriginal place now known as The Block, was unlikely and un-expected for the (sub)urbanised non-Aboriginal majority of Austra-lians. Aboriginal people seemed to belong 'back then' or 'out there' in the 'outback' and, as such, were largely forgotten in the cities, at that time.

Beyond the visibility of whiteness

Toni Morrison has produced a corpus of scholarship on 'race' and racialization in the United States. In addition to her academic writings (for example, Morrison 1992), I have found informative disruptions to the hegemonic ethnicity of whiteness in her thought-provoking novels. One portrayal that I will never forget, was the opening sequence in the novel *Paradise* (Morrison 1998), which detailed the excruciatingly brutal murder of a group of unconventional women at the hands of the 'upstanding' men of a town. Regardless of the book's opening line 'They shoot the white girl first', *white* ethnicity (as it is understood in the wider US context and in my Australian understanding as well) was largely absent from this particular story. The story was set in one of the many African American towns pioneered in isolation during the long marches from slavery and as I read this opening account, a seemingly familiar scenario of oppression suddenly yielded something quite unexpected, and unfamiliar, to me. First, the perpetrators of the crime were not 'white', ethnically speaking, which was what I had expected (from reading stories about brutal oppression of minority groups). Second, and the real bombshell for me, was that the 'white girl' did not fit my understanding of the term 'white' – she was (I believe) an Indigenous ('Native') North American. My particular (mis)reading of Morrison's story had occurred because of my ignorance (gasp!). I was not aware of much of the historical setting for this story but I had assumed (albeit, with just a little prompting) the necessary presence of a particular form of 'white' ethnicity.[24] My positioned reading of *Paradise* – and I am fairly sure that I am not the only one who made such blundering assumptions – revealed to me that another kind of whiteness was in operation even in the *absence* of the (expected) ethnicity within the story. I had inadvertently naturalized, centralized and, in this case, incorrectly perceived a universal ethnicity. It was the process of my defaulting to such a categorization, as well as the assumption itself, that constituted a *processural* form of whiteness and alerted me to the need to critically interrogate the embedded, discursive and yet arbitrary ways of whiteness.

One of Morrison's earlier novels provided a conceptual link. In *The Bluest Eye* (Morrison 1970), an 'ugly' child carried a double burden of the historical geographies of oppression from without, and brutality from within the group to which she ostensibly belonged. The portrayal was of the child's descent into a pact to end the torture of her disposition through the acquisition of an impossible (and probably imagined) pair of blue eyes. For this child, designations of ugliness included the darkness of her skin tone, hair and eyes, and her poverty. This text alluded to the particularities of oppression, the psychoses that resulted from the impacts of nightmare pathologies of bigotry where memberships and non-memberships were indicated through certain designations. *The Bluest Eye* is a story of status through designations of beauty, rather than 'race'.[25] Yet, at the same time, Morrison demonstrated that although formations of bigotry are not the sole domain of any particular group (ethnic/'racialized' or otherwise), an aspect commonly associated with (some) 'white' people – blue eyes – was *the* aspiration in this story.

Morrison's richly detailed accounts are often of worlds that are largely external or *other to* the worlds of ('Anglo') 'white' people (and I am categorized in this way). And yet these stories do acknowledge the realms of dominance within which they, in the US context, must exist. Of equal importance, Morrison's stories invalidate one of the more common tasks of whiteness studies, which is to make 'whiteness' visible, ethnically speaking. As Morrison has carefully demonstrated, centring (white) ethnicity, even in the study of whiteness, can be highly ethnocentric. As Bonnett (2000, 120) has remarked, [whiteness is already ' "the natural order of things" that has structured its representation[s]...thus the meaning and formation of whiteness are taken for granted.' So the general law, grand narrative or ethnicity of whiteness is not what needs to be exposed. Rather, it is the particularities of whiteness, within the full context of circumstances that have enabled a range of empowerment possibilities (Wiegman 1999). Problematizing its hegemonic ethnic status is one way to open out the concept of whiteness to new and more useful critiques.

To assist in moving the whiteness debate beyond its characteristic reductionism, I have adopted the usage of more overtly applied postcolonial/imperial perspectives. North American whiteness studies, as a sub-discipline has tended to neglect Indigeneity[26] because of its particular preoccupations with the more overt racialized binaries of post-slavery. Apart from the historical moment of segregation, mentioned earlier, I am aware, and Morrison demonstrates this well, that Indigenous/'native' North Americans are sometimes considered to be 'white' regardless of their variously racialized positionings (in Australia, Aboriginal people are usually associated with 'black'). However, in critical race studies (in Geography and elsewhere), postcolonial theorists and researchers have raised the need to consider the trajectories of colonialism experienced by Indigenous peoples, *everywhere*, up to the present (King 1992, Peters 1998). The inclusion of post-colonial/imperial perspectives, in studies of whiteness, also holds the potential to liberate research from the trap of (mis)representation of the 'other'. To exemplify this positioning, the next section situates post-colonial/imperial perspectives (with an emphasis on the post-colonial) within a study of location-specific processes of whiteness.

Cities of Neo-colonial Whiteness

There is a widespread conception, that goes beyond Australia, that urbanization of Indigenous peoples is a bad idea. For some, it defies 'nature' – they belong outside urbanized modernity. For others, it is a dangerous final stage in a long process of assimilation. Urbanization does have its pros and cons, and in the increasingly urbanizing context of Canada, Bonita Lawrence (2004) has identified the loss of status and identity for 'native' peoples[27] through urbanization. Others have provided a somewhat different perspective. Indigenous peoples are 'as culturally creative and adaptable as anyone else' according to Jeffrey Sissons (2005, 63), who posited that urbanization does not necessarily strip cultural distinctiveness. Furthermore, he identified that the formation of networks, afforded in urban settings – particularly before the days of advanced telecommunications – have been pivotal in the rise of Indigenous politics. This was the case in the US, with the formation of the World Council of Indigenous Peoples, and with the advent of *Indianismo* in Bolivia, Colombia and Ecuador, which Sissons (2005, 75) has described as a 'coherent nationalist framework'. In Australia, the rise of Indigenous politics was largely an urban expression. The coalescence of a range of Indigenous peoples, from many cultural groups, occurred originally because of issues of dispossession, rather than political desire. Another element, the availability of otherwise unwanted urban areas to inhabit (a pull factor), combined with rural and 'outback' pressures – of shrinking 'Country' and increasing unemployment (push factors) resulted in urban Indigenous settlements. These new unifications, which may have placed kin-based identities in the background, certainly resulted in a critical mass that *then* mobilized. This resulted in the badges of 'Aboriginal Australia', such as the Aboriginal flag, that were foundational in the overall politicization of Indigenous Australians. Urban spaces that are largely undesirable to the majority have long provided cheap lodgings for disadvantaged groups. And as Bonnett has identified (2002, 362–365) these settlements of the 'other' have:

> ... been structured around and mapped on to a tradition of spatialized light/dark dualism that can be seen emerging in the 19th century. Although the notion of 'white cities' is no longer viable in such contexts, the symbolic role and function of confined, highly demarcated zones of 'non-whiteness' within metropolitan societies continue to be significant ... the association of the urban, modernity and whiteness is a deep-rooted tradition[.]

The largely unexpected formalization of an urban Indigenous settlement in inner Sydney in 1973, certainly unsettled the predominantly non-Aboriginal nation (Anderson 1993a & b). It represented the authorized demarcation of a 'non-white' zone within a society that believed it was immune to the need for 'racial segregation' (within its cities, at least). Since then, and particularly with the onset of gentrification in the late 1990s, in what did become one of the most overtly segregated corners of urban Australia, 'black' (Indigenous and surrounded) and 'white' (non-Indigenous and surrounding) spaces butt up against each other in increasing juxtaposition. The mass media portray The Block as a 'failed human experiment', and as 'a ghetto' – as Harlem-like (Shaw 2000). In response to its existence, an array of strategies and projects aimed at consolidating and reinforcing 'white' space find expression. By opening these fields, using the kinds of analyses now commonplace in Human Geography,[28] the following vignettes demonstrate the allocation of ethnicity, consequent racialization, and the subtler performances of whiteness enabled in these localized settings. These examples reveal also that processes such as the racialization of particular spaces may be context specific, but these forms of whiteness also reverberate within the wider context of neo-colonial Australian politics.

Particular whiteness

As a long-term resident of inner Sydney, I have observed a range of machinations associated with the largely unwanted presence of Indigeneity, and an ongoing desire to destabilize the existence of The Block.

One of my earliest observations of whiteness occurred in 1995, when I participated in an initiative to 'hand over' the provision of community welfare training to local Aboriginal people. A new 'Community Welfare' course offered training for Aboriginal welfare workers at Eora[29]

College, which is within easy walking distance of The Block. During the final stages of completion of their qualifications, the first intake of students suddenly found that they were required to attend the main (and mainstream) campus in another part of the city. Funding had been cut from Eora because of the 'replication of existing provision' within the Technical and Further Education (TAFE) college system.

As the only Aboriginal TAFE College specified by ethnicity in the State of New South Wales, Eora stands alone. The other colleges do not carry a tag of ethnicity, which reflects the societal values of the (apparently) non-ethnicized majority of citizens. Within the TAFE college system, the non-ethnicity of dominant norms has been institutionalized with Eora being the only 'ethnicized' exception. In the case of the Community Welfare initiative, the small group of students, many of whom had experienced adverse impacts of the ('white') welfare system in the past,[30] and who were dedicated to the Eora project, gradually discontinued their studies. They found it too wearing to continually justify their politics, their hardships and, at times, their existence in the mainstream (mostly non-Aboriginal) course.[31]

My next and most striking observation of whiteness complicated my earlier observations of ethnicized spaces. Even though distinctly racialized spaces operated in my neighbourhood, the range of 'ethnicities' present is always more complex than a simple rendering of 'black' and 'white'. The population that utilizes 'white' space includes individuals from a myriad of 'ethnic' backgrounds.

Redfern Railway Station on Lawson Street, Redfern, has a high flow of pedestrian traffic entering and leaving at peak times. On a daily basis I see the pedestrians streaming along one side of the street, while the other side (The Block side) remains virtually empty. Commuters heading to and from the nearby university deliberately avoid walking on The Block side of the street.[32] An invisible line has long determined the 'safe' side for the diverse group of commuters, and a 'no-go zone', which is for the others and the recent addition of barricades has enforced this zoning). Regardless of who inhabited the space, ethnically speaking, the non-Block side of the street becomes a *temporary* space of whiteness in juxtaposition with the other side. Moving away from the starkness of highly defined 'black' and 'white' spaces, where the space of whiteness absorbs all non-Aboriginal[33] ethnicities, whiteness (like blackness)[34] *appears* to fade, but the default ethnicity of 'whiteness' remains intact. These simple observations of the ascription of ethnicity to whiteness, have helped me to identify some of its covert operations.

Unearthing historical geographies of whiteness

The paradoxical presence of The Block has elicited a range of responses from the wider non-Aboriginal community. A sense of siege and defensiveness has built on a history of colonially-based class relations that were compounded by the invasion and seizure of Indigenous lands in 1788. Before The Block's formalization, impoverished residents of what was then the blighted part of the city (which had declined with industrialization) battled to save their suburb but lost half of it to university expansion. The second threat to existing residents occurred with the proposal to establish an Aboriginal settlement in this already besieged place.

Formalization of The Block occurred at a time of widespread ignorance of Aboriginal people. For most non-Aboriginal Australians, Indigenous people lived traditionally, somewhere in 'the outback'. Regarded with suspicion, a bourgeoning urban 'black politics' was associated with the Aboriginal settlement. Referenced with US politics, civil unrest seemed inevitable as this part of Sydney became a focal point for Indigenous activism, along with Aboriginal migration to the city. Aboriginal people became more 'visible', as a 'radical' part of the wave of social reform movements of the time (Mickler 1998). The 1967 referendum[35] had consolidated Aboriginal identity nationally, as did establishment of the Aboriginal 'Tent Embassy'[36] in 1972 (Attwood 1989), which indicated a new stage of Aboriginal politics in Australia. Meanwhile, inner Sydney residents had come face-to-face with a new, unknown and, for many non-Indigenous people, 'frightening' presence. Some of these Redfern residents, including those who were migrants from parts of Europe, embarked on another struggle to protect their embattled enclave from 'invasion'.

More than 30 years after formalization of The Block, urban Indigeneity remains unacceptable for many non-Aboriginal Australians. Many residents, who fought against the establishment of The Block, have succumbed to the forces of gentrification but the new incumbents often harbour familiar resentments. The image of danger that is continually percolating has elicited a frustration with the unwanted 'other' that is familiar in urban landscapes around the world (for example, Gregory 1993). In Australia, this has combined with a more widely felt, historically embedded, and romanticized ideal of egalitarianism. This notion of a 'fair go' for all, ironically in this instance, seemed to be intolerant of racial segregation. Although introduced in the 1970s (1966 in South Australia, see Bulbeck 1993), anti-discrimination legislation has not addressed unspoken and unwritten segregation. At that time, the non-Aboriginal majority of Australians lived in suburbs that, although designated by class, were not associated with common understandings of segregation. The presence of a 'racially' segregated place, an Aboriginal-only place, incited widespread resentment,[37] which is 'an emotion...[and] a power with its own material and discursive logic' (Solomon, in McCarthy et al. 1997, 234), that continues to this day (Hage 1998, Mickler 1998).

The Block is widely perceived to be a privilege – just like the Aboriginal-specific Community Welfare course at Eora – created for Aboriginal people. In response, the mobilization and activation of urban settlement has had devastating effects on the Indigenous community. *Cities of Whiteness* teases out processes that reveal the elusiveness and multiplicity of whiteness as it encounters Indigeneity, in the local, at the level of the state and, on occasion, internationally.

Although dominated by the relatively new US-based sub-discipline of whiteness studies, I have suggested that identification of the *geographies* of whiteness have the capacity to go beyond the current themes particularly the identification of 'whiteness' as an ethnicity. Rather than identifying whiteness as a (powerful) zone on a segregation map, or a description of (ethnic) difference, this book supports Bonnett's (1996, 97) assertion that whiteness is 'temporally and spatially contingent and fluid'.

This chapter has introduced some of the ways that I began a process of unravelling whiteness. I have sketched a trajectory, from my initial suspicions about whiteness – as observed in a very indirect, but nonetheless definite racialization event experienced by Aboriginal participants in a TAFE course – through to an ongoing research agenda that seeks to unveil the strengths, and slipperiness, of whiteness.

From around my home in inner Sydney, I have observed how whiteness performs and consolidates itself in this place through a range of pressures exerted directly and indirectly on an urban Aboriginal community. Socio-cultural shifts have reinforced an ongoing construction of an overt and racialized binary, which is continually re-imagined through processes of urban change. Its newest expressions, of select heritage sensibilities and new urbanities, have evolved to the deliberate detriment of the 'Black Capital of Australia' (as it was originally founded), which continues to shakily clutch its stake at the heart of Australia's first and most global city. Away from this immediacy, whiteness resumes its subtler capacities.

The presence of an impoverished urban Aboriginal community in inner Sydney has provided a catalyst for the observation of locally specific performances of whiteness to be captured, and documented. Such observations, of the specificities of this inner-city place, reveal the historical geographies of colonialism, and its cultures, that continually replicate and reinforce normative values of racialization, more generally. The local context provides specific processural details, such as the turf-protection that occurs with the designations and protection of heritages, that are historically embedded in colonial structures of privilege. In the city of Sydney, processes of gentrification and urban transformation more generally, which are driven by a range of external (including global) as well as local forces, are tethered to select pasts. As the next chapters will reveal, beliefs about heritage and entitlement to place replicate a colonial project of excluding the Indigenous other. Parallel to these selective identifications of heritage, Sydney's 'Manhattaniza-tion', which is detailed in Chapter IV speaks of another set of pasts, presents, and futures that are whitewashed in the neo-/post-colonial gentrifying city.

Notes

1 Also spelt Cadigal.
2 In Australia, the terms 'Aboriginal' and 'Indigenous' are often interchange-able. They are used in this way here. Indigenous refers to those who identify as such rather than any biological and/or cultural determinant (though it is acknowledged that some Indigenous peoples prefer these). Common attri-butes of Indigeneity/Aboriginality include (knowledge of) kinship ties, and familial histories of dispossession through (ongoing) colonisation. I capitalise 'Indigenous' and 'Aboriginal', and all associated versions of these terms in acknowledgment of Australian Aboriginal protocols.
3 Somewhat confusingly, The Block straddles Redfern and the tiny suburb of Darlington (Figure 1.2) but most Australians associate it with Redfern.
4 As Australian Bureau of Statistics social maps (www.abs.gov.au) demon-strate, Australia *is* highly segregated by class (income) and 'ethnicity', regardless of popular understandings of egalitarianism. However, urban segregation is not as overtly delineated as in the United States, for instance. There are few places that are designated as 'ghettos', or places that the majority fear to go.
5 'Trans' refers to 'across' in 'transdisciplinarity' and in this context means that the study of whiteness borrows from across the disciplines. 'Inter', as in 'interdisciplinary scholarship' refers to 'between', which means that the study of whiteness is not owned but shared among the disciplines.

6   African American writers, such as Toni Morrison, have challenged this assumption. I discuss examples from two Morrison novels later in this chapter.

7   According to the most recent Australian Bureau of Statistics estimates, 'the highest proportion of Aboriginal and Torres Strait Islander peoples...liv[e] in major cities (30%)...20% liv[e] in areas classified as inner regional, 23% in outer regional, 9% in remote and 18% in very remote areas'. Australian Bureau of Statistics, 4713.0 Population Characteristics, Aboriginal and Torres Strait Islander Australians, 30 October 2003, updated 18 March 2005 (www.abs.gov.au/ausstats/).

8   The Block is also dubbed 'Redfern' or 'Eveleigh Street', but I will refer to it as it is most commonly known.

9   Referring to its general and widest sense (see Endnote 3 and Figure 1.2).

10 Positivism was a view that aligned geographical (and other) inquiry/ies with scientific 'proof' through quantification.

11 For example, Dear (2000, 1) reacted to the criticisms with the comment that postmodern urbanism 'risks being regarded as hopelessly faddish, already obsolete, or terminally indecisive. Such criticisms are usually the product of hostile or lazy minds'.

12 Such as the ethnic/racial categorizations depicted in social atlases (see Cynthia A. Brewer and Trudy A. Suchan, 2001, *Mapping the Census 2000: The Geography of U.S. Diversity*, ESRI Press, Redlands California). Maps of US Census data includes sets titled 'White', 'Black or African American', 'American Indian and Alaska Native', 'Asian', 'Native Hawaiian and Other Pacific Islander', 'Two or More Races', 'Hispanic or Latino Origin', and the intriguing 'non-race' or 'ethnic' category of 'White, Not Hispanic or Latino Origin'.

13 As used by Hage (1998). There is a popular assumption is that 'Aussies' are generally 'Anglo', with a few 'migrant' and Aboriginal exceptions. Levels of migrancy, with the highest intake from Britain and/or New Zealand (which, of course, includes non-'Anglos' but the majority have long been migrants of 'Anglo' heritage), disrupt this assumption.

14 Geographers include Alastair Bonnet, who championed the critique of whiteness from within Geography. Other critiques include publications by Dwyer and Jones (2000), who attempted to unsettled the 'socio-spatial epistemology' of whiteness, and Winders (2003) who identified the 'power framework' of whiteness.

15 'Revolutionary and counter-revolutionary theory in Geography and the problem of ghetto formation', from *Social Justice and the City*, 1973.

16 *The Black Inner City as Frontier Outpost*, 1974.

17 A new, cross-disciplinary form of whiteness research has recently emerged (with a new journal and association, both titled *Australian Critical Race and Whiteness Association* (http://www.acrawsa.org.au/index.php/item/244) that focuses specifically of questions of Indigeneity (see Anderson 2002, and McKay 1999).

18 'Native title' is the name given by the High Court to Indigenous property rights recognized by the court in the *Mabo* judgment (3 June 1992). The Mabo judgment overthrew the legal fiction of *terra nullius* – that the lands of Australia had belonged to no one when the British arrived in 1788 (http://www.atsic.gov.au/issues/land/native_title/Default.asp).

19 Although the police involved were exonerated after the Abernethy Inquest, the result was controversial and the Aboriginal community (and others) were outraged. At a public meeting, held on 8 October 2005, the new Premier of New South Wales, Morris Iemma, was called to re-open the case. As yet, this has not occurred.

20 This 'riot' was picked up by the international media, and reported in Al Jazeera, Onenews New Zealand and IOL South Africa on 16 February 2004, BBC News and New York Times on 17 February 2004. In the UK, The Guardian also ran a series of articles and commentaries over several days. Commentaries and debates were ongoing in Australian newspapers, such as the *Sydney Morning Herald*, the *Australian,* the *Age,* and television coverage was widespread.

21 In December 2005, a widely reported series of 'race riots' started on Sydney's Cronulla Beach, when 'Anglo' youth allegedly decided to reclaim the beach from other ethnicities (particularly 'Lebanese' people).

22 Pulido (2002, 45) has estimated that the discipline of Geography is 'over-whelmingly ethnically white' (over 90%).

23 Interviews with Horton on 'race' at http://www.media-diversity.org/articles_ publications/Hue%20and%20Cry%20on%20'Whiteness%20Studies'.htm and on 'whiteness' at http://www.pbs.org/race/000_About/002_04-back-ground-02-04.htm.

24 See Bonnett (2000) for discussion of non-Anglo identifications of white-ness, throughout history.

25 It is noted however that part of the designation of beauty was racialised (darkness of skin tone).

26 And, as I am very aware, and Morrison has demonstrated, Indigenous North Americans are sometimes 'white' regardless of their variously racia-lised positionings.

27 In 1901, only 5.1% of Canada's Aboriginal people lived in cities, and by 1951 only 6.7% had urbanised, but by 2001, almost half of Canada's Aboriginal people lived in urban areas (Peters 2005). According to Evelyn Peters (2005, 346), there is also a characteristic that complicates the urbanisation of First Nations peoples – they tend to 'churn', that is they move between cities and homelands, and back again. This is also common in Australia.

28 The research documented here used a multiple-methods approach that drew on qualitative, interpretative, and quantitative sources, to generate data and a writing-as-analysis approach to constructing the research narrative.

29 Named after the Eora peoples who, before colonization, inhabited the coastal area where the inner city of Sydney now stands (Kohen 2000).

Eora is the only Aboriginal campus of Technical and Further Education (TAFE) Colleges in the state of New South Wales.

30   A high proportion had personal experiences of the welfare initiatives that resulted in the 'stolen generations' of Aboriginal children (see note 1 in Introduction).

31   These accounts were reported to me personally.

32   The University of Sydney provides information on the safest route to walk to Redfern Railway Station (www.security.usyd.edu.au/basic/unis. html#map) and provides a free security bus service to and from Redfern Railway Station that runs at night. Walking across Lawson Street with a colleague, a group of students were overheard debating the topic of crossing the road and facing the dangers of walking on the other side. They did not cross over.

33   Many local Aboriginal people suffer the ills of chronic poverty. When they cross to the 'safe' side of Lawson Street, they are generally viewed as out-of-place, threatening and invasive.

34   Aboriginal people are far from a homogeneous group but on the 'black' side of Lawson Street, all become one feared and generally unacceptable group, regardless of who they are (such as welfare or health workers) and why they are there.

35   On 28 May 1967 a federal referendum gave the Commonwealth constitutional powers to legislate on Aboriginal matters by amending Section 51 (xxvi) of the Constitution which gave only the States such powers. The 1967 referendum authorized the deletion of Section 127 of the Australian Constitution so that Indigenous people could be counted in the census (Mickler 1998, p 121). Voting in Australia is compulsory.

36   The 'Aboriginal Embassy', set up on the lawns of Federal Parliament in Canberra, was widely demonised and eventually 'violently removed by police under a new law... introduced by the Liberal government' (Mickler 1998, p 139).

37   As do other 'enclaves' of 'ethnic' concentration such as the suburb of Cabramatta in Sydney's west (see Dunn 1998).

2
(Post)colonial Sydney

To flesh-out empirically my proposed (re)conceptualization of whiteness that moves beyond ethnicity, this chapter sets the scene for a series of urban processes that have occurred, and are occurring in the inner city of Sydney, Australia. The first section, *From Dangerous to Endangered City*, details a context for recounting some of the stories of whiteness that follow the potted history of the city's settlement, and provides a detailed background for more recent urban processes to be discussed in the following chapters.

The exile of Aboriginal people from the Sydney area began in 1788, with colonization and settlement. Nearly two centuries later, the act of handing land 'back' to Aboriginal Australia within the largest and most international Australian city,[1] remains a paradox for many non-Aboriginal people. Indigeneity, for many, seems almost antithetical to the notion of progressive globalization – it is largely associated with isolated and disconnected 'outback' locations, which are not far removed from the harshness of nature. The dishevelled state of the inner city Aboriginal settlement in the neighbourhood of Redfern is therefore not that unexpected; for many it appears to be fulfilling its fate by moving towards extinction. And yet, The Block persists, well beyond wider expectations and the shrill discourses of its decline (cf Beauregard 1993) perpetuated by the media, police and new occupants of the gentrifying city. This chapter tells the story of the birth of The Block, and how the part of Sydney in which it sits became a site of struggle that continues to this day. Additionally, I consider the production and maintenance of values about Indigeneity – that stereotype and stigmatize. I have found that these values are grounded within colonial and neo-colonial historical geographies. The events detailed in this chapter have set a foundation for ongoing processes of racialization of the Aboriginal community in inner Sydney, and Aboriginal Australians more generally. Some very subtle alliances have formed to maintain a project of marginalization and exclusion; some have evolved as part of the changing urban landscape. Such alignments are also part of a larger story – a wider conceptualization – of neo-colonial whiteness.

From Dangerous to Endangered City

Like many other cities in the world, Sydney has had its own experiences of 'gentrification'.[2] The march of the 'middle classes', from the suburbs 'back' to the inner city, began in the 1960s in this city. Before that, neglected inner-city neighbourhoods provided cheap spaces for the poor to inhabit. The most disadvantaged were Indigenous Australians – Aboriginal peoples – who began migrating to the 'big smoke' in the 1930s. There were many reasons for migration to cities at this time, but settlement in most undesirable areas occurred because Aboriginal people had few options. Inner Sydney provided refuge for the most dispossessed, hungry and threatened. Apart from the ravages of invasion, the shootings and poisonings, Aboriginal peoples were also forcibly moved from their traditional lands – from Country – to reserves, from the late 1890s (Reynolds 1996). Massacres of isolated Aboriginal groups were still occurring in the 1930s (and probably later), and disease, malnutrition and poverty were rife on the reserves (Bulbeck 1993). The general belief among non-Aboriginal Australians, corroborated by government policies and scholarship, was that the demise of these 'primitive' peoples, and their cultures, was inevitable. As with so many Indigenous, First World, 'Native' and other Aboriginal peoples around the world, it was widely understood that this was a dying race; it was not expected to survive the upheavals of modernity.

Migrations to the city of Sydney meant the possibility of employment as well as sanctuary. After decades of the breakdown of family networks through forced separations by 'protective' governmental schemes (Anderson 1993a & b), many also travelled to Sydney in the hope of finding information about their kinfolk. 'Redfern' (in this case, the Darlington/Redfern and Chippendale area)[3] was the focal point of such migrations and Redfern Railway Station was the place to alight after a long train journey from 'the bush'. The entrance to Eveleigh Street, now the main street of The Block, is just across the road from Redfern Railway Station.

Apart from housing poor Aboriginal people, Redfern's symbolism rests on two quite distinct attributes. For Indigenous Australians, Redfern holds political importance as the site of historical struggle, and a focal point for the birth of a range of Aboriginal services and, it is a meeting place for otherwise displaced Aboriginal peoples (Anderson 1998). Second, it is widely recognized for its troubles. In the late 1960s and early 1970s, an overt politics of place had emerged (Anon. 1974, Anderson 1993a & b) and Redfern became synonymous with Aboriginal struggle. The formalization of The Block, which provided a plot of land and housing rights for Aboriginal people in the city of the original colonial settlement (and place of invasion), was a turning point in Australia's history. In a nation that has yet to come to terms with the existence of Indigeneity, in a place originally declared to be *Terra Nullius*, for which no formal acknowledgement or government apology has been forthcoming, The Block attracts a particular kind of attention. As the popular media in Australia know all too well, for imaginings of what is 'wrong' with Aboriginality, simply cast a spotlight on The Block. However, rather than highlighting a history of the politics of Indigenous struggles for rights, and its role as a magnet for the most dispossessed peoples, criticism of The Block comes from all sides – even from within (discussed later). The Block is held up as an example of Aboriginal failure – a 'failed human experiment' as it has often been described – and its fate forever hangs in the balance. At the place of British invasion and subsequent colonization, The Block is a discomforting reminder that in a nation proud of its multiculturalism, issues of 'race relations' run deep. Representations of Sydney's 'Redfern Riots', which gained international media attention, attest to this.[4] The Block is subject to heavy surveillance and the image projected perpetuates the kinds of stereotypes that feed imaginaries of Aboriginal inferiority which are frighteningly unevolved from the days when tall ships entered that revered harbour.

The dangerous city

Not far south of Sydney Harbour, the neighbourhoods of Darlington, Redfern and Chippendale[5] were built on the Blackwattle Swamp, in the mid-1800s. There are few records of the pre-Victorian era (named after Queen Victoria of Australia, and the rest of the British Empire of course) but the legacy of those times – of decorative terraced houses – remains (McInness 1967). A second notable stage was a descent into a long-term stasis known as the 'slum era' that gripped the whole of inner Sydney. From the turn of the twentieth century, the dirt and grime of industrialization, overcrowding and inadequate infrastructure exacerbated impoverished living conditions (Fitzgerald 1987). A third stage in the evolution of 'Redfern' (as I'll now refer to Darlington, Redfern and Chippendale unless referring more directly to each place) occurred during the era of suburbanization after World War II, which resulted in further deterioration of the neglected inner city. Disinvestment and decline continued throughout the 1900s, until a fourth stage when the history of this pocket of city space diverged from the rest of inner Sydney. The movement 'back', with gentrification, began in the 1960s and 1970s, but was highly location-specific for reasons that I will return to later in this chapter.

Industrialization and 'slum' landscapes

Throughout the nineteenth century, most Sydney dwellers lived in Victorian or 'Federation' style terrace, or row, houses (Figure 2.1) that were 'not too far from Sydney Cove' (Frost 1992, 193). Home was usually close to work in the small, but swiftly growing city. By the *fin de siecle*, Sydney's ubiquitous terrace houses were increasingly regarded

Figure 2.1   Terrace or Row Houses in Sydney

as 'an inherently bad form of housing that fostered slum attitudes, crime and immorality' (Fitzgerald and Keating 1991, 80). The catastrophic epidemic of rat-borne bubonic plague of 1900, and the subsequent smaller outbreaks in 1902, 1907 and 1921–2 (Curson 1985), affirmed the increasing undesirability of inner-city worlds. The poorest parts of Sydney were Ultimo and Pyrmont, just to the west of the Central Business District (CBD), Woolloomooloo to the east, and Surry Hills, Erskineville, Alexandria, Waterloo and Redfern, to the south. These neighbourhoods industrialized first and any remaining houses became 'slums' (Howe 1994). As these areas deteriorated environmentally, and socially, the former homes of the ruling classes – the large terrace houses, which are located on main thoroughfares – were converted to boarding houses (Fitzgerald 1987). As the Redfern area began to industrialize in the 1920s, small wooden terraces made way for factories, warehouses and breweries. Although Chippendale (adjacent to the northerly boundary of The Block) changed dramatically due to industrialization, Darlington (to the south of Chippendale) retained much of its predominantly brick housing stock until the era of university expansion, in the 1960s.

Urban blight and suburbanization

The transformation of inner cities, from places of general settlement to zones of ill-repute and disregard, occurred hand-in-hand with processes of suburbanization (which are summarized in Chapter I, page 00). Meanwhile, the groups that migrated to Sydney from war-torn Europe after World War II, found that they had moved to a nation which was firmly divided by class and religion. The two main classes were the predominantly working-class Irish Catholic population,[6] and the more affluent White Anglo-Saxon Protestants (WASPs). By 1947, Australia had recorded its lowest proportion of overseas-born since colonization (Bulbeck 1993). Then, a sudden flood of new migrants arrived who were not English-speakers, and who brought with them a set of very different cultural practices. This registered as 'culture shock ... [for the then] highly homogenous anglicized white population' (Bulbeck 1993, 131). The concentrations of these 'first generation migrants' were widely regarded as unassimilated into mainstream Australian society. Rather than following the nation's new home-ownership ideal this group of migrants rented housing in the inner city. These residents of the then unwanted and generally ignored inner city were largely powerless against what was to come.

The endangered city

By 1960, the tiny neighbourhood of Darlington, within the Redfern area, was under siege. Residents battled to save their suburb from complete subsumption, by The University of Sydney (herein, the University). Then, in the 1970s, a second stage of besiegement occurred with the proposal to formalize the Aboriginal presence as an Aboriginal settlement.

The non-suburban location, and the mix of industry, housing, ethnic diversity and poverty, had contributed to the undesirable reputation of the Redfern area. It was also by-passed by the usual forces of gentrification, until other options had diminished. But a monumental threat did occur in that late 1950s. In 1957, the University appointed a Select Committee to develop a plan for its future.[7] This committee outlined the proposal for expansion, and university planners turned their sights on the deteriorated, largely undesirable, suburb of Redfern. The result was the large-scale expansion of the University into the tiny neighbourhood of Darlington. This caused major social upheaval with the displacement of whole sections of residential streets. Homeowners, as well as renters, were displaced. Owner-occupiers were given assistance to buy elsewhere after receiving an amount near, or just above, the property value set by the Valuer General's Department (a NSW government department)[8]. But the majority of residents were 'protected tenants in the terms of the Landlord and Tenant Act 1939' (McInnes 1967, 353), and the University assisted some of these residents to move to other rental properties. In the ensuing mêlée, the University offered leases for the housing it had purchased during this expansion process. This meant that some displaced protected tenants moved into the houses of their unprotected neighbours. According to McInnes (1967), there were incidents where tenants had moved in to these houses, spent their scarce savings decorating them after assurances that they would be settled for five or six years, only to find that they had to move again 18 months later because of changes to university planning. Residential reaction was understandably 'adverse' and after heavy lobbying, 'certain aldermen [councillors] of the Sydney City Council' initiated a process to save Darlington from complete subsumption by the University (Conybeare et al. 1990, 15). The final plan of expansion known as the 'University Extension Area' was duly reduced. The remaining half of Darlington was saved (UEA, see Box 2.1).

According to the University's development report, planning for expansion had been hampered by the 'attendant problems of finance, procurement, over-spreading and *antagonism from citizens in affected neighbourhoods*' (Conybeare et al. 1990, 22 emphasis added). Since that time, the residents of Darlington, and the Redfern area more generally, have consistently met any further developments proposed by the University, with scrutiny and protest. In 1997, the University and local residents came to an arrangement for 'community consultation' via a 'Community Committee' (according to a local former 'committee' member and local resident and business owner, Interview 17 October 1997) as part of any development process. The residents remain suspicious – the history, and memory, of those earlier expansionist days has fed an ongoing community distrust of any incursions into the area.

Box 2.1 University expansion

According to the University of Sydney's Strategy Plan (Conybeare, Morrison and Partners 1990), the UEA for Darlington consisted, in part, of land originally granted in 1835 by Governor Bourke. After WWII, 50 acres of the land to the South of City Road in Darlington, were earmarked for the new University of Technology (now the University of New South Wales, UNSW). When UNSW was located in Kensington, the Darlington land was set aside under the Cumberland County Plan to meet the expansion needs of the University of Sydney, and the adjacent Royal Prince Alfred Hospital.

The 1958 UEA proposal was that the University would subsume Darlington and in the 1960s, the University began the process by subsuming 438 residences (some were shops with residences). Various other shops, 47 factories, 5 pubs, a dance hall, a bottle yard, a branch of the Commonwealth Bank, a junk yard, the Royal New South Wales Institution for Deaf and Blind Children (which became the University's Department of Geography), Darlington Public School, St Michael's Hostel for Catholic girls, Darlington Town Hall, Darlington Post Office, an electricity sub-station and two children's playgrounds were also swallowed (McInnes 1967, 348, 349). A 'Special Uses' zoning for schools, and the University, was applied to the Darlington area on the 11 May 1958 but the exact boundaries remained uncertain. The boundaries of this zone were finally set in 1969 when the University was allocated an extra 9 acres (Proceedings of Council, 1959, Sydney City Council), which meant that parts of Darlington would remain unaffected. Darlington Road, Codrington Street, Abercrombie Street and Golden Grove Street now bound the University's Darlington Campus. The South Sydney Council Local Environment Plan, 1996, confirms these current boundaries.

Loss of Darlington identity and threats to 'heritage'

The struggle for territory in the Redfern area is reflected in the near loss of the suburb *and* the name of 'Darlington'. A main arterial road, Cleveland Street, separates Darlington and Chippendale but since the University's expansion, both suburbs are routinely referred to as 'Chippendale'. Having lost territory to the University, and then to The Block, and sharing a postcode with Chippendale, Darlington only just exists. However, in recent years, Darlington has re-emerged as a separate entity through the efforts of real estate agents, and others with vested interests in awakening its identity. For example, a newspaper headline: 'No-Go for taxis in Chippendale' (*Sydney Morning Herald*, 27 September, 1997, 3), referred to the stigmatized 'no-go' zone around The Block, which is actually Darlington. But now, 'Darlington' is distinguishable from 'Redfern' and Chippendale – particularly in the real-estate business of property sales. Somewhat ironically, Darlington now benefits from its association with the University of Sydney – the oldest university in the country, which was modelled on the designs of Oxford and Cambridge,[9] in the United Kingdom. This historically linked identity has worked well with the recent surge in 'heritage' appreciation, and the rise of 'heritage housing', in inner Sydney.

In the late 1990s, a battle over heritage and taste was waged in Darlington. Another university, the University of Technology Sydney (UTS), had built student housing on the site of the somewhat lamented 'Miss Muffett's Jam Making Factory'. Local residents' distrust of the monolith, the University of Sydney and its expansionist agenda, applied in this case. The UTS development – a modern student-housing building – became quickly and thoroughly regarded as the image of *anti*-heritage. Although the old factory had been demolished, and the new structure built without much public commentary, at a public meeting some months later, local opinion was loud and clear. The local council, South Sydney City Council (SSCC), called the meeting to consider the fate of another site, which sits adjacent to the UTS building (14 June 1997, Darlington). At this meeting, members of the recently formed 'Resident Action Group' (RAG)[10] mobilized a campaign to protect the site from a similar fate. The site, which consisted of an old factory, was earmarked for conversion into apartments. The overwhelmingly emotion expressed at the public meeting was that the UTS development represented a benchmark for anti-heritage 'tastelessness', that should not be replicated. As one angry resident stated: 'We want more sympathetic [development]... with 2 storey terraces. When's the 3rd, 4th, 5th and 6th [student housing] building [going ahead]? Is Council serious about...heritage?' (South Sydney City Council Public Meeting, 14 June 1997).

UTS had become another (in this instance, aesthetic) invader and therefore a threat to Darlington's heritage. The residents of inner Sydney have become adept at mobilizing to protect 'heritage', and I will return to this in more detail in the next chapter.

The resented 'return' of Indigeneity

Another significant and highly controversial event contributed to the sense of siege and threat, and consequent resident mobilizations in the Redfern area. In the early 1970s, another patch of Darlington was earmarked to for an urban Aboriginal settlement, which became The Block. This event set in train a set of battles, of resentment and turf protection that continues to this day. By the 1970s, the residents of the already diminished suburb of Darlington were well used to mobilizing for a fight when the then Federal Labor Government granted funds to purchase that now infamous site for Aboriginal housing.

The allocation of land and Victorian terrace houses for Aboriginal people unsettled the already somewhat destabilized and wary non-Aboriginal community in Darlington (Anderson 1993). It occurred at a time of widespread ignorance about Aboriginal people. For most non-Aboriginal Australians, Indigenous people were considered to be somewhat primitive, and dwellers of the 'outback', the barren desert landscapes of the centre of the continent. In general, non-Aboriginal people were surprised by the groundswell of 'black (Aboriginal) politics'. The urban Aboriginal presence was regarded with suspicion – they were considered to be 'half-caste troublemakers' unlike their 'full-blood' cousins who seemed happy living a natural and traditional existence. As Australian 'black politics' strengthened, and increasingly associated with the US model and civil unrest, there was even more disquiet.

As the point of Aboriginal migrations to the city, Redfern also became the focus for activism throughout the late 1960s and early 1970s. Aboriginal people became more 'visible' to non-Aboriginal society. Aboriginal people gained Australian citizenship in 1965. Then, in the 1967 referendum[11] Australians voted to reform Indigenous policy: for Indigenous peoples to counted in the census, and Indigenous matters to become national (Federal), rather than State-based concerns. The Aboriginal 'movement' accelerated, and the 'Tent Embassy'[12] was erected on Australia Day, 26 January 1972 (Attwood 1989). This initiated a new stage of Aboriginal politics, and militancy, in Australia. Back in Redfern, local (non-Aboriginal) residents had come face-to-face with the new and confronting 'black politics'. This Aboriginal mobilization was treated with suspicion – it was viewed to be too 'radical' (Mickler 1998, 115). Then, in 1973, the newly elected federal Labor government made provision for the purchase of land and housing for the Aboriginal cause. But this was far from well received and appeared to be a show of support for the wrong group. The proposed hand-over of part of Darlington/Redfern to Aboriginal people, who had suddenly become highly visible with their demands, galvanized non-Aboriginal residents into a new politics. It seemed that *their* representative government – a Labor government – has supported the frightening and highly politicizing 'other' instead of its traditional 'working classes' constituents. At that point, a form of powerful and highly protective enclave consciousness was borne.

Although Aboriginal people had lived in the city for a long time, the formalization of The Block meant that the Indigenous presence could no longer be denied, or ignored. Regardless of its legitimacy, its presence, which has been widely regarded as creating a 'no-go zone' or ghetto, was (and is) definitely not considered to be appropriate settlement in an Australian city. For many, The Block was doomed from the start and its predicted decline is long overdue.

Securing Whiteness in the Paradoxical City

Because Aboriginal people were so thoroughly exiled from the place of colonization, their presence in the city remains unexpected – an unsettling paradox for many non-Indigenous Australians. Fear and paradox nourish each other, and feed into what John Gabriel (1998) has termed a 'whitewash' effect. Later in this chapter, I will refer to this as 'whitewish', meaning the desire for a particular kind of exclusionary gentrification that permeates the Redfern area.

The next section outlines some of the methods used to ensure that The Block remains paradoxical, which has become part of a strategy of securing whiteness. I argue that because the presence of The Block is a paradox for many – the settlement of 'outback' Indigenous people in the city does not really make sense for much of Australian society – it is therefore improbable. This has led to a belief that its demise is only logical. The ongoing practice of monitoring the Aboriginal community by police, residents, and the mass media, has perpetuated an ongoing discourse of The Block's decline (cf Beauregard 1993). The area has a reputation – it is dangerous and a feared place. It is a 'fearscape' (Davis 1998) perpetuated by media, and other representations. These have continually cast local non-Aboriginal constituents as victims of the criminally inclined Aboriginal perpetrator. The extension to this logic is that the decline of The Block will bring the desired closure to this chapter of besiegement in inner Sydney. The promise is for engagement in urban life that has done away with a fearful perpetrator. The under-lying and mostly unspoken promise[13] is for unfettered gentrification once The Block is dismantled.

Maintaining the paradoxical status of urban Indigeneity

From the outset, The Block has been 'out of place' (Cresswell 1996), which is reflected in itscontinually reinforced and reiterated paradoxical status. I now turn to the building of the portrayal of The Block as always unsuitable, as *not belonging*. Through careful monitoring and media reportage, the status of The Block as an unacceptable settlement, has evolved. Not only was The Block not wanted, it was resented because it was seen as a 'gift' for the unworthy. The 'myth of [Aboriginal] privilege' (Mickler 1998) is part of a more general era of backlash against minority groups. The belief that 'they' get too much – through affirmative action or other policies, welfare provisions and so on – has conflated with widely held stereotypes about Aboriginality. Such stereotyping, which in this case is that the acquisition of an undeserved privilege (houses and land in the city) has also been squandered, has exacerbated The Block's status as untenable.

Systems of surveillance
Since its inception, The Block has been a magnet for surveillance. From Foucault's observations of Bentham's 'Panopticon' (1977), through to Davis's (1990) 'fortress city' and Soja's corporate 'citadel' and 'paranoid architecture' (1990, 1996), the politics of surveillance has been well recognized. Surveillance represents a capacity to regulate, restrain and contain. More than mere observation, surveillance is a sophisticated and multifaceted process that can result in domination through the defence of space, or result in the loss of public spaces (Newman 1972 on 'defensible space', Urry 1990 on 'gaze' and public space, Keith 1993 on surveillance/policing, Hillier and McManus 1994, Morgan 1994, on surveillance in Australian cities). Surveillance technologies have become

part of everyday life and are generally expected in shopping malls and other privatized 'public' places. Surveillance in spaces of consumption, such as shopping precincts, may provide comfort for those who just *go about their business*, but it is determinations of what are *un*acceptable uses of such spaces that are often inflexible and discriminatory. A controversial example is the 'shopping citadel' of 'City Walk' in Los Angeles, in California. This is an exclusive, gated 'leisure enclave' for shopping (Beckett 1994). The unauthorized businesses and activities of poverty, which include collecting empty bottles, asking for spare change, or simply sitting in one place for too long, have been deemed unacceptable. The uses of such spaces are defined, in part, by what is *not* allowed and the rigid norms of acceptable use are maintained through careful planning and security measures, including 'close circuit television' (CCTV) and patrols by security guards. Such realms of surveillance have begged questions about who fits in, and who does not, and the rationales for exclusion (Cresswell 1996). In the case of groups of young men being 'moved on' by police or security guards in shopping malls, age and gender are relevant determinants, as well as behaviour or class (as is particularly the case with obvious poverty). 'Race' too, is often a determinant of non-acceptance (Keith 1996).

In recent considerations of racialized exclusion, John Gabriel (1998, 12) has linked the politics of surveillance with a politics of 'whiteness': 'Surveillance . . . is an integral aspect of whiteness . . . [and it] cannot be understood outside of discursive, regulatory and technological means at its disposal to position itself through others.' In Sydney, CCTV networks constantly monitor urban spaces that have been identified as crime 'hotspots', and the area around Redfern Station is heavily surveilled. In inner Sydney, surveillance is tied to the 'fortressing'. Surveillance is perpetuated in loops of 'security' measures, which circulate between residents, who watch and report to police (and, at times, gain media attention, which is discussed in more detail in the next chapter). There are links between residents, businesses and private security guards, and police. Many homes and business premises have 'back-to-base' alarm systems, which are linked to private security firms, and local police stations. The police use video surveillance as well as regular foot, car and helicopter patrols. Redfern Railway Station has CCTV surveillance and specialist railway police keep watch. The police and the local council (which was South Sydney City Council, but is now part of the City of Sydney) report details of their activities and concerns back to the residents through leaflets, newsletters and public meetings. Private security firms promote their products, and *the need* for their products, heavily, in the Redfern area. Local resident action groups respond quickly to incidents of crime, particularly if they are linked to The Block (considered in detail in the next chapter). Institutions, such as the federal government's former Aboriginal and Torres Strait Islander Commission[14] (ATSIC), and the NSW Departments of Aboriginal Affairs and Housing, that have provided (and retracted) funds for social services, are also stakeholders in 'keeping an eye' on The Block. As a local Aboriginal spokesperson remarked: 'We are talking about the most surveilled area in the country...Redfern police have had continuous electronic surveillance in operation at The Block [since mid-1997]' (Lyall Munro, 31 January 1999, Australian Associated Press, Reuters).

The Block is certainly heavily scrutinized. Such scrutiny has created a skewed version that continually confirms its problematic status. The perpetuation of fear that comes with observable high levels of surveillance technologies serves to reinforce the need for such technologies, and the fear.

Fearscape production

From the 1960s onwards, inner Sydney fearscapes have gradually faded with the march of gentrification. Conversely, in the Redfern area, the advent of The Block sharpened existing fears of the already dangerous and dishevelled inner city. The discourse of inner-urban decline telescoped onto Redfern, and the neighbourhood around The Block retained its fearscape status until the 1990s, when other Sydney gentrification options had dwindled.

The poor image of The Block, has been exacerbated by the drug-related incidents depicted in the media (Box 2.2). The ongoing reportage of such incidents reinforces popular opinion about The Block, and stereotypes about Aboriginal people, more generally. One Darlington resident described The Block as a 'time bomb' (interview 16, 29 April, 1998). This resident felt that it was a place where incidents could escalate, and explode, at any time. More recent and highly overblown reports of 'Sydney's Worst Riots' confirm this viewpoint (*SMH*, 16 February 2004, 1).[15] Then on 'Australia/Survival/Invasion Day', 26 January 2006, I was unable to drive home using my usual route. The 'Survival Day' celebrations on The Block had just finished and a fight had erupted on a street corner. In response to this altercation, consisting of two men fighting, streets were barricaded with lights flashing and police, who were ready for a riot. I was told that I would have to find another way, and circumnavigated the whole area to get home.

Box 2.2 'Drug headlines'

'Tackling The Block: Police move in on Redfern's drugs and crime,'
'Heroin feeds wave of crime', *SMH* 18 January 1997
'Zero tolerance as police blitz streets', 'Drugs, despair and derelict houses chip away at The Block', *SMH* 16 January 1997
'Say no to drugs for the next generation,' *Chippo Politics* 1, 9, December 1997
'Local Drug Fear', *South Sydney Bulletin*, 9 September 1998
'Residents united in drug paper,' *South Sydney Bulletin*, no date
'Minister: It's wrong,' *Sun Herald*, 31 January 1999 (re-white boy using heroin in Aboriginal place)
'Heroin shot – Aboriginal spokesman says he smells a rat', *The Australian*, 31 January 1999
'Redfern – a bogus war on drugs', *The Australian*, 1 February 1999
'Experts condemn needle shutdown', *SMH*, 1 February 1999
'Heroin site no surprise to minister', *The Australian*, 2 February 1999
'Needle give-away – Box of 100 syringes left in lane for addicts', *Daily Telegraph*, 2 February 1999
'Redfern closure wrong', *Canberra Times*, 2 February 1999
'Carr calls summit,' *Sun Herald*, 7 February 1999[a]
'Residents fear Redfern clinic will be needle exchange,' *SMH*, 17 February, 1999
'Carr defend needle exchange program', *The Australian*, 20 May, 1999 ('concerns over the needle exchange program in Redfern precipitated the drug summit')

[a] 'Carr' refers to the former Premier of New South Wales, Mr Robert Carr

The media seize upon any incidents that occur between police and residents of The Block. The *Sydney Morning Herald*, Sydney's most popular broadsheet newspaper, maintains its vigilance. Any incident on The Block risks being dubbed 'a riot' and making newspaper headlines. This has been a routine occurrence since the 1970s. Typical examples of riot reportage include '[Redfern] riot case for trial' (*SMH* 2 December 1989, p4) and 'It was Redfern, but it was just like Tobruk' (*SMH*, 7 August 1989, 3), which compared Redfern (The Block) 'riots' to war. 'Sorry we couldn't stop them, say police...car stolen by rioters in Redfern' (*SMH* 3 January 1990, 1), and 'Call for tougher penalties after police attacked [during a riot] in Redfern' (*SMH* 27 January 1997). Such

---

Box 2.3   Policing The Block

'Police accused of brutality in Redfern arrest of Aborigine', *SMH*, 20 October 1989, 3
'Arrests follow police action in Redfern', *SMH*, 22 January 1990, 6
'Row looms on inquiry into raid at Redfern', *SMH*, 23 February, 1990, 18
'Police raid racist, says churchman', *SMH*, 2 March 1990, 2
'Redfern police raid 'racist': report condemns excessive force', *SMH*, 21 May 1990, 1

---

reports confirm the unruly status of the area. On occasion, police too come under media scrutiny. In 1989 to 1990 there was a spate of reports on policing The Block. Emotive terms such as 'riots' and the policing response, known as 'raids' have peppered media reportage on relations between police and Aboriginal people in Redfern.

Police are variously represented. On occasion, they are supported, and considered to be victims in the realm of dysfunction associated with The Block. Reports, such as those listed above (Box 2.3) do attract the criticism that the media is 'police bashing' (Interview, 27 April 1998). The police have been accused of a variety of misdemeanours, such as 'underpolicing', 'overpolicing' and racism (Cunneen 1990) but they are defended; they are seen as misunderstood and misrepresented. A resident, who lives very close to The Block, expressed increasing agitation (and anger)[16] about media representations of police:

> Very often...sections of the media do...the old sacred cow treatment of the Kooris[17]. [They report that] all these dirty [whites/police]...come in and flog these poor blacks. It's just ridiculous, because blacks are flogging whites and vice versa...its the poor police who've got to deal with these filthy, ugly human scenarios...it's just like 'Cop It Sweet'. Remember 'Cop It Sweet'?...It wasn't only cop[s copping] it sweet, everybody's copping it sweet. Two nights later [after 'Cop It Sweet' was broadcast] my wife came home and got bloody mugged. (Interview, 15 November 1997)

According to this resident, police and local non-Aboriginal people have to 'cop it sweet'.[18] (In using the phrase 'cop it sweet' the resident was making explicit reference to a television documentary on police racism, 'Cop It Sweet', see Brockie 1991). This resident (above quote) was quite forthcoming, and angry with what was perceived as media bias against police. The view was that the media continually privilege the Aboriginal 'side' of the story to the detriment of non-Aboriginal people, and largely non-Aboriginal police, who all were victimized by living, and working, with the Aboriginal presence.

Another resident, who had lived in the Redfern area for many decades, also mentioned the 'Cop It Sweet' documentary:

> Sometimes I think they [the media] blow it out of proportion, living around here, but sometimes [we]...cop it sweet. Yes, we've seen that [person] in the car, and I reckon it was a pinched [stolen] car...but they don't care...what I mean is they've [media/Aboriginal people?] just got a set against the coppers [police] and that's it...They [Aboriginal people] reckon they own all the land and...they never paid a penny for it...Our fathers went and fought for the land, I don't know whether their [Aboriginal] fathers did but I know mine did to make it what it is now, but they don't respect that. The media...sometimes they do blow it out of proportion, but then again they sometimes do tell the truth, and sometimes, they don't know the truth. We who are living around here, we do. (Interview, 27 April 1998)

This long-term resident felt that there was a contradiction – non-Aboriginal people and police were the victims (of media reportage, and Aboriginal crime). For this resident the memory of 'fathers' defending the nation during the world wars sets a benchmark for a value system of respect. Although it was acknowledged that Aboriginal 'fathers' might have fought for 'the land' (Australia), during the World Wars[19], there was no acknowledgement for what had been lost by Indigenous Australians before those wars. For this resident, it was the soldiers who were being shamed because the land they defended and, by inference, its laws and value systems, were no longer respected. The paradox is that it is 'we' – who for this resident were the non-Aboriginal residents, particularly the older ('working-class') people – and police, who were besieged, and the media failed to recognize this 'truth'. Several police officers I spoke to also felt that the media misrepresent policing in the area, and that this was particularly so in the 'Cop It Sweet' television documentary.[20]

A local newsletter, *Chippo Politics*[20] (November 1999, 4, 9, page 8), which was part of the alternative press at the time, carried a very different line on policing:

> There are a couple of issues that remain on the agenda...one...is the issue of police/black relations. There seems to be enough evidence that the police are still resorting to their old tactic of [using] a sledgehammer to crack a walnut.

maintain the 'fearscape' that exists around Redfern Railway Station. In 2005, Redfern Police moved into one of the TNT towers, which loom over The Block, the railway station, and the street. The fearscape is also of particular interest to new residents.

Infectious fear

The feelings of fear, the fear of loss of safety, and the feelings of besiegement that circle around this fearscape, can be simply categorized as 'race fear'.[23] For non-Aboriginal residents, and others, fear and distrust are focused on the racialized 'other', and this is continually reinforced through media depictions. For those who feel besieged by the presence of The Block, the 'danger' script perpetuated by the media (and others) has provided a handy scapegoat, an outlet for all the tensions of living in the inner city. It is easy to blame The Block for everything that is wrong with the city.

To compound an already fraught situation, this ongoing racialization of all that is problematic, reinforces the 'black–white' binary. It *forces* a certain visibility of 'whiteness' (Dyer 1988). As discussed in the previous chapter, the group that is usually ethnically neutral is suddenly rendered visibly 'white' because of the contrast with the racialized 'black' group (Thiel 1991). The flipside to the discomfort of being exposed, ethnically, is that whiteness appears pristine in contrast to blackness. Within this racialized binary, whiteness is continually cast as innocent, and as the (fear filled) victim. The Aboriginal community of The Block carries a stigma that is familiar to other racialized groups, where 'race' is associated with drugs, dirt and disorder[24] (Douglas 1966, Kristeva 1982, 1991, Sibley 1995), and crime. Another example is Cabramatta, in western Sydney, which is stigmatized as an 'Asian' drug-crime hotspot (Dunn 1998). In these situations, whiteness is always at risk of being sullied. Around The Block, vigilance in surveillance, and defence of space, remain a priority for non-Aboriginal residents and business people, and for those who pass through the area.

Securing the status of paradox: discourses of Indigenous decline

During the late 1990s, the status of The Block declined further as the fearscape image escalated. A new discourse of The Block's decline helped to embed its paradoxical status in the city. Once sanctioned as something that cannot belong, its demise seemed inescapable. The discourse of decline pointed to a closure in this chapter of Aboriginal Australian settlement in the city of Sydney. 'For once, Aborigines [sic] can't blame White Australia . . . [e]xhausted and in disgrace, Redfern is an Aboriginal failure, and they know it' (Journalist's commentary, *SMH* 18 January 1997).

In what can only be described as 'blaming the victim', residents on The Block are continually blamed for their circumstances. In the 1990s headlines included 'Drugs, despair and derelict houses chip away at The Block' (SMH 16 January 1997, 6), as the discourse of The Block's decline reached a crescendo. In 1997, the media reported that the Aboriginal Housing Company (sometimes called the Aboriginal Housing Corporation, and formerly The Aboriginal Housing Cooperative), who administer The Block and act as landlord, had announced that the site was to be transformed. The plan was to be rid of the problem by demolishing and redeveloping the site, possibly into a 'Black Chinatown', with an Aboriginal cultural centre (*SMH* 22 March 1997). The media, real-estate agents and others, seized on this announcement to redevelop the site, and housing prices escalated. Astute property developers turned their eyes to the neglected warehouses around the area, which were ripe for 'conversion' to apartments (which I discuss in detail in Chapter 4).

An important point to mention is that it is not solely the domain of non-Aboriginal Australia to criticize The Block. The unquestioned belief that The Block has failed, that it has succumbed to drugs and crime *because* of its Aboriginality, gained further legitimacy through the contribution of Aboriginal voices. Comments by high-profile Aboriginal spokespersons such as the late Dr Kumantjayi Perkins were widely reported. Concerning conditions on The Block, Dr Perkins allegedly stated that 'we've got to kick black arse over this' (*SMH*, 18th January 1997, 33, 44). According to newspaper report, the Chief Executive Officer of the Aboriginal Housing Company has also stated his desire to 'finish it' (The Block)[25]. Aboriginal interventions usually highlight issues such as disadvantage or, as in the Dr Perkins' commentary (*SMH* 18 January 1997), community dysfunction. Although the media jumps at the hint of Aboriginal criticism[26] of anything Aboriginal, it rarely reports on the many efforts made, the support for The Block, or the sheer tenacity behind its survival. Even sympathetic reports end up as stories of despair. Of course, the discourse of decline is much stronger when sanctioned by Aboriginal voices. The broader picture of disadvantage and dispossession that are the context for current problems, and the successes of many programs such as those provided by local agencies, are simply not newsworthy.

Whitewash

Earlier in this chapter, I mentioned the term 'whitewash', and will now return to this idea and its relevance to events in inner Sydney. Whitewash, according to John Gabriel (1998) is a process of cultural bleaching whereby the racialization of others naturalizes the dominance of whiteness. Processes of whitewash around The Block have occurred with urban transformation, with gentrification, which includes the recent push by apartment developers into the local area (Figure 1.2). The strategies of surveillance of The Block and anti-Block resident action that have increased with gentrification, also contribute to whitewash. The racializing impacts of such activities are not obvious; they are sidelined by the concerns about controlling drugs and crime, and the potential spread of 'black' space.

The media's capacity to contribute to such whitewashing processes was exemplified well in the aforementioned story about Dr Perkins' alleged sentiment of blame (*SMH* 18 January 1997). Because an Aboriginal person, particularly an Elder, and high-profile spokesperson, cast the blame of some arbitrary measure of failure directly at The Block and its Aboriginal residents, the popular reading will always be clear. Aboriginal people are to blame for circumstances on The Block because an Aboriginal person said so. Yet in other cases, Aboriginal voices are far from heard, in fact they are silenced. What is absent from this reading is that a particular emphasis was seized upon because it was sensational and newsworthy. This choice of *which* story to report occurred elsewhere, in the hallowed halls of news production. At that time, the story sought was designed to meet an agenda, which was a verification of the (ongoing) discourse of The Block's decline. The next example was a little more explicit. An announcement by the Aboriginal Housing Corporation – to change the status of The Block from residential to a 'Black Chinatown' – aroused the interest of the ABC TV 'Four Corners' unit. With a reputation for its in-depth, investigative journalism, particularly after the exposé of police racism in the 'Cop It Sweet' documentary, this report promised to introduce the 'bigger picture' of the pressures on this tiny urban Aboriginal community, to the media spotlight. The promise was for the definitive story, presented from the *inside* by a highly respected and award-winning ABC journalist.

The journalist, and the production team, spent time with the people interviewed and gained a level of trust (Jackson 1997). There were many moments of intimate and sympathetic communication between the residents and the journalist. However, regardless of the sensitive portrayals of individual stories, the documentary makers could find no conclusions to draw other than the already 'known' situation of despair. So rather than a sensitive, in-depth investigation that dealt with the issues at hand, this documentary simply corroborated the usual sensationalist representations of drugs and drug use, poverty (replete with footage of rats scuttling about, as they do around inner city lanes after dark), devastated lives and hopelessness. This 'gritty realist' investigation was unable to escape the lived realities and, in the end, served as just another version of the usual discourse of decline about The Block. The style of the report echoed what McCarthy et al. (1997, 237) have identified in relation to the 'wave of black cinema in the USA' (with directors such as Spike Lee), which has

> a kind of documentary accuracy...[whereby] vendors of chic realism recycle a reality code...[in] the elaboration of a hierarchy of discourses – the fabrication and consolidation of secular common sense.... [These films use a] 'constructed reality code of 'being there'. (McCarthy et al. 1997, 237)

Although the Four Corners documentary was clearly not 'black cinema', the production, a predominantly non-Aboriginal project, was a representation *of* a place, filmed in a 'being there' style. Unlike 'black cinema', the secular common sense in the Four Corners documentary was the sensible middle-class view about the unacceptability of dirt, and the horrors of poverty. This outsider-view was superimposed on the people of The Block who are already well aware of their circumstances. Regardless of the intentions of the documentary makers, which I am sure were well meaning and the intention was to provide a sense of 'being there' (which was still rather voyeuristic to say the least), they were unable to see beyond the shock-worthy images of dysfunction, and self-perpetuating despair. The context, the external realities and the 'bigger picture', as well as the other stories of everyday lives, and struggles against dispossession, unemployment, lost families and racism, continue to be sidelined to the seduction of providing images from Sydney's own 'war zone'.

Box 2.5   Other stories about The Block

A film was made by a small local welfare agency, The Settlement, titled: *Fight for Blood* in 2000. *Fight for Blood* was based on the lives of various individuals from *The Block* (past and present), including the ghost of a deceased drug user. *Fight for Blood* won three film awards including the prestigious Griffith Film Festival Prize of 2000 ($10,000). Later, a documentary produced by the Indigenous Cultural Affairs Magazine (ICAM) at SBS TV titled: 'Substance Misuse Part 2' (22 March 2001) featured a sensitive portrayal of the difficult lives of drug misusers on The Block. It dealt with issues of unemployment and despair without subscribing to a discourse of The Block's decline.

The international press have also found newsworthy stories from The Block. The prominent US broadsheet, the *New York Times*, carried an article headline 'Enclave Reflects Aborigines' Plight: Developers covet a zone the Koori find crucial to unity' on 9 February 1997. Unfortunately, this promising headline was misleading. The opening sentence read: 'In the shadow of gleaming skyscrapers three subway stops from the CBD, young men . . . sell heroin openly'. The 'plight' of Aboriginal residents, at the hands of property developers, was overshadowed by this exaggerated portrayal of criminality. The interesting story (worthy of the headline) of the coveting developers and the pressures they place on The Block must have ended up on the editorial cutting-room floor. These sorts of media accounts have charted what is now a highly drawn-out process of The Block's demise and yet, the *actual* demise is yet to happen.

If we believe what the newspapers tell us, The Block has been *almost* gone for decades, as illustrated in the headlines listed in Box 2.6. There have been some false-starts to its demise, and the inevitable may be slower than expected, but the media picture remains clear: The Block is heading towards extinction regardless of the scepticism among local non-Aboriginal residents, who remain unsure about what was going to happen. Those who live near The Block have their own theories about its failure and the following quotes are particularly explicit about failure:

I think over the past 5 or 6 years it's been really the drug problem. It [The Block] should never have been up there ... it's a terrible mistake and the ones who got away and moved out into the suburbs are probably leading much better lives, away from there, it's just too concentrated ... but also it's all the country people, the blow ins ... they come in from the country, they think ... we can go a bit wild here while we're visiting the relatives. (Interview, 29 April, 1998)

Box 2.6   Survival or Demise of The Block

1996
'Redfern Survival? The future of Koori culture in the city', special issue *City Hub*, 25 January 1996
'Eveleigh St land rights claim: emotion runs high as residents battle for their block', *SMH* 3 March 1996
'Black Santa sleighs The Block a final time', *SMH* 23 December 1996

1997
'Its days are numbered', and 'The Block's days are numbered', *SMH*, 18 January 1997,
'Blacks vow to fight demolition of Block', *SMH*, 1 February 1997
'$6m to raze The Block', *Daily Telegraph*, 4 February 1997,
'Perkins linked to new plan for Block', *SMH*, 22 February 1997
'Block Out', *The Australian Magazine*, 14–15 June, 1997
'Block saved from builders', *Daily Telegraph*, 14 August 1997
'Block to get the chop? Not whilst houses are occupied, says Mayor Vic Smith,' *Sydney Bulletin*, 19 August 1997
'Block bulldozed: Eveleigh St reduced to rubble', *Daily Telegraph*, 18 December 1997
'Enclave Reflects Aborigines' Plight: Developers covet a zone the Koori find crucial to unity', *New York Times*, 9 February 1997
'Report puts blacks back on The Block', *Daily Telegraph*, 1997

1998
'"I'm staying put"' says Elder', *Sydney Bulletin*, 27 January 1998

1999
'Spotlight will be on Aborigines [during Olympic Games],' *SMH* 18 June 1999
'Bulldozed!', *Sydney City Hub*, 4, 38, Cover story, 6 May 1999
'The Block's last line of defence, save her 40 great-grandchildren,' *SMH*, 50, 483, 21 June 1999

to the demolition and relocation of The Block's residents after a survey indicated 40 of the 52 families wanted to leave' (*SMH* 21 June 1998).[28] The cost of demolishing The Block is regularly recycled, and inflames resentment about taxpayers spending money on ('unworthy') Aboriginal issues. Examples of this recycling include: '$6m to raze 'The Block' (*Daily Telegraph*, 4 February 1997, 1,2), 'ATSIC seeks $6m to demolish ghetto' (the *Australian* 4 February 1997, 3) and 'ATSIC puts up $6m for Redfern facelift' (*SMH* 4 February 1997, 4). The aforementioned '52 families', who apparently wanted to leave The Block, was based on the details of a survey commissioned in 1995 by The NSW Government Department of Aboriginal Affairs, and funded by the 'Building Better Cities' initiative. What the newspaper articles failed to mention, however, was that the Redfern Aboriginal Coalition (RAC), who represent tenants and other residents on The Block (as well as administer employment programs), disputed the results. The RAC found gross inadequacies in the interview process. Most notably, only the official leaseholders were interviewed, but the majority of the population living on The Block did not hold leases. Leaseholders were made a raft of promises about the provision of better housing, and many moved out to take up such offers (by the Aboriginal Housing Company). The glaring misrepresentation in much of the reportage was that the residents (apart from the Elder, mentioned previously) *had already moved away* regardless of the obvious occupation of houses. One article stated:

> [s]he [The Elder] will only leave her home of 20 years when she is convinced the area will remain in Aboriginal hands, and that *the remaining local Aboriginal community*... has a home. 'When they're safe... and I find a place to my liking, after winter, then I might move,' she [allegedly] said. (*SMH*, 21 June 1999, *my emphasis*)

Obviously, there was some confusion about the population on The Block and about whether its demise was imminent or not, but every few months we (the news-consuming public) seem to be reminded that The Block is going, or nearly gone. We are led to believe that drugs, crime, declining standards of living, government machinations and so on, have taken their final toll, yet again.

The aforementioned article (*SMH*, 21 June 1999) did provide a shift in emphasis, from sensationalism about crime and living conditions and The Block's gradual implosion, to identifying the final marker of its passing. Presented through the eyes of the 'last' survivor, this article echoed other stories of the so-called passing of Aboriginal people.

The death of Truganini in 1876, for example, was significant because she was reported to be the last Tasmanian Aboriginal person[29]. Historian Henry Reynolds (1999, 13) noted that

> in both popular and specialist literature [it was agreed that] the Tasmanians had been a uniquely primitive people ... living fossils, representatives of the 'old Stone Age'... the implication for mainland Aboriginal people was clear. They too would tread the downward path [towards extinction].

The descendants of Tasmania's Indigenous peoples, who recently gained an apology and promise of recompense for the 'stolen generations' from the Tasmanian State Government,[30] must be non-plussed about the ongoing belief that they have passed into extinction. The focus on the last surviving resident of The Block momentarily quietened the usual noisy barrage of narratives about crime and fear. It seemed inevitable that The Block would pass (into extinction) once the remaining resident had gone. There was a strange reverence in the reportage. But, this was suddenly disrupted, when, at the end of the 'Good Bye Aunty' article, the Aboriginal Housing Company's chairperson, at that time, stated: 'We're trying to improve the place [The Block] and it will be for Aboriginal people and it will be housing' (*SMH* 21 June 1999). There may have been displacement of Aboriginal tenancy on The Block, which heralded its passing but the Aboriginal *residential* presence would, it appeared, remain. There are two versions of what was occurring on The Block. There are those who discursively dismantle it, and those who continue to fight for housing. The latter includes the voices of the Redfern Aboriginal Corporation (RAC) and the Redfern Residents for Reconciliation (RRR). In 1998, a high-profile Aboriginal spokesperson, Sol Bellear, stated that those who think otherwise (i.e. that The Block will not be a place for Aboriginal housing), are 'in for a surprise' (interview, 15 September 1998). As local non-Block residents know all too well, the remaining few official tenants on The Block may have moved away, but some do return. The Block's status as a meeting-place, and the dire need for shelter in a run-down house, in a tent, or sleeping out around a campfire, ensures an ongoing Aboriginal presence on the site.[31] The reality is that The Block remains the 'Black Capital' for some of the most dispossessed Aboriginal Australians. It continues to attract people who often have nowhere else to go, or who return after trying to live in mainstream suburban Sydney. Many stay and rent or purchase houses nearby. Others live on as squatters. For many who have lost connection with their cultural origins, The Block is home, and for others it is the place to find out about their origins – their 'Country' (Anderson 1999).

Whitewish – the promise of gentrification

The discourses of decline about The Block have reinforced a general belief in the inevitability of its demise. The discourses are a part of the wishful thinking of a post-Block fantasy. Another aspect of the post-Block fantasy has occurred through the very material processes of inner city transformation. Gentrification is gradually changing the Darlington/ Redfern and Chippendale area. As gentrification has progressed, many have speculated and purchased properties, which remain lower in cost with proximity to The Block. Such speculative investments have largely relied on the prospect of an escalation in property values. Such speculations have tended to be based on belief in the forthcoming post-Block era, supported by a plethora of (media) 'evidence'. Such beliefs have relied on the rhetoric of 'change', and the location has certainly changed over the past decade. Investment in apartment developments has soared (detailed in Chapter 4), and house prices have escalated. The usual trappings of gentrification, the cafés, art galleries, furniture shops and so on, have appeared. Yet, the most expected event, the demise of The Block, remains a speculative gamble. It is far from fact.

According to Real Estate Institute figures (Suburb Snapshot, www.smh.com.au), between March 1998 and March 1999 median house prices in the Darlington and Chippendale postcode area (2008) rose by 3.5 per cent in a market of general decline in house prices (for example Paddington house prices *dropped* by 5.3 per cent). The rise in prices in the Redfern area can be attributed to its 'affordability' for home-buyers or 'low entry' investors, as compared to other inner Sydney areas. The sudden surge in apartment prices in the Redfern area, and nearby, was dominated by warehouse or 'loft' conversions. Apartment prices escalated by 18.3 per cent from March 1998 to March 1999 in Redfern, but in Paddington (the 'hallmark of gentrification') apartment prices dropped slightly. The media seized upon these statistics to promote the inevitability of a gentrification scenario but the assumption made about the kind of gentrification that has occurred, and the sort peddled by marketeers, needs to be viewed with a critical eye.

The common misconception perpetuated by real-estate agents, and associated media discourses, is that gentrification processes are standardized. The reality is that not all inner-city areas in Sydney will 'do a Paddington' as was claimed for Redfern, by the media ('At $508,000

Redfern May do a Paddington', *SMH*, 31 August 1996, 3). Gentrification does not always fit in with such predictions. The assumption that gentrification will follow the Paddington lead has stimulated investment, which in turn has stimulated further investment in 'frontier' areas, such as Redfern. In the context of Sydney's gentrification, areas such as Paddington and Woollahra are generally thought of as being at a 'mature' stage of gentrification. By contrast, the Redfern area is classified, particularly by real-estate selling agents, as being in the *early* stages of gentrification, with the promise of higher capital gains on property investments. Like Glebe or Newtown, gentrification in the Redfern area may peak at a level that is different from the established benchmarks. The attainable house prices, and associated cultural and taste capitals, such as 'retro' versus 'Victorian' aesthetics, may never be the same. To promote such uniformity in gentrification is illusory. Setting other areas against the benchmark of Paddington and Woollahra is often misleading, particularly for gentrification speculators. It is, however, good for the business of buying and selling property but, the imaginary of a mature gentrification scenario for the Redfern area, like that of Paddington, cannot include the presence of The Block. So, real-estate agents, and the media, have promoted the assumption that a new post-Block era is just around the corner. For investors the new era for this last frontier of gentrification will blossom once cleansed of the final impediment.

Conclusions

The first part of this chapter outlined the paradox of urban Indigeneity in inner Sydney. It detailed some of the changes that have occurred in the neighbourhoods near The Block, and other parts of inner Sydney. I outlined some of the history of Darlington's besiegement, the establishment of a community spirit of suspicion, and the threat to territory that has come with the arrival of The Block.

In the second part of this chapter, I documented the emergence and maintenance of a value system that stereotypes and stigmatises Indigeneity in the post-colonial city. The use of fear and surveillance has perpetuated a scenario of The Block's decline. I have outlined the ways that local narratives generally support, but sometimes question, this idea of impending decline.

Overall, The Block continues to be discursively framed as a dangerous place. Of specific concern to this chapter was the practice of monitoring the Aboriginal community, by police, residents and the mass media. Such monitoring has assisted in perpetuating the discourse of The Block's decline, and its construction as an ever-evolving 'fearscape'. The non-Aboriginal community around The Block continues to be cast as the victim of the criminally-inclined Aboriginal perpetrator. Discourses of decline about The Block not only support the probability of a hopeful closure to this chapter of besiegement in inner Sydney, they actively pursue this end through a process of whitewash. The promise – the white*wish* – is for urban lifestyles without fear for person or property, *and* unfettered gentrification (and the associated capital gains once The Block has gone). In the meantime, The Block continues to inspire the scripting of this part of Sydney as 'dangerous' (but with potential). This chapter has considered such scripting and its reinforcement of a racialized binary, where 'black' is associated with night, dark lanes and Aboriginal people, and 'white' with daytime, well-lit streets and gentrifiers. Whiteness is imagined to be the highly organized and vigilant protector that, above all else, retains control of a situation that seems always poised to spiral out-of-control. Not unlike the colonial encounters of the past the fantasy of whiteness, in this instance, is founded on and justified by the fear of the (black) 'other'.

For me, as a local resident, I know that there is a heroin problem (which is probably no worse than anywhere else in the inner city), and the level of sickness bring ambulances screaming into the area. There are also fires that summon fire engines[32] into its narrow streets. I also know that the less eventful days lived on and around The Block are much more common. I can report that I have lived, for more than a decade, in a neighbourhood with occasional dramas, most of which are noisy moments – a heated verbal exchange now and then, and the overblown police responses. Ultimately I remain continually amazed at just how quiet my place really is, particularly when I compare it to my previous inner city addresses, and The Block is just around the corner. The rare incursions into my lifestyle do not cause me concern – I live in the city, and like most, it is a city of difference.

Notes

1   Sydney is the most powerful city in the region, and is considered to be a 'fledgling' global city by world standards.
2   I remain somewhat ambivalent about the notion of gentrification and its blanket generalizations – particularly its current popularization by governments, real estate marketeers and property developers, gleefully embracing (and profiting from) 'creative' urban renewal, as a salve to urban blight.

3   'Redfern' refers to the suburb of Redfern, the Darlington/Redfern and Chippendale area and also The Block, particularly in media accounts.

4   For example, stories of Sydney's 'Redfern Riots' were reported in *Al Jazeera*, 16 February 2004; *Onenews New Zealand*, 16 February 2004; *BBC News*, 17 February 2004; *IOL South Africa*, 16 February 2004; *The New York Times*, 17 February 2004; and ongoing commentaries for several week after in *The Guardian*, UK). Later that year, the Macquarie Fields 'riots' were similarly reported, and were followed in December 2006 with 'race riots' in beach suburbs.

5   Even more confusing than the name Redfern, The Block actually straddles Darlington and Redfern, and borders Chippendale, so the whole are is often dubbed 'Redfern', particularly by Aboriginal people, because of the close proximity of Redfern Railway station.

6   See Ignatiev (1995) for discussion of the 'whitening' of Irish Catholics in the USA, and Hickman and Walter (1995) for the British context.

7   In 1961, enrolments reached 12,527, which increased to over 18,000 by 1985. This indicated that there had been an accurate forecast of growth (Report on the development of the University Site. 1961–1985, 1985).

8   Such figures are usually well below market value.

9   http://www.usyd.edu.au/alumni/faqs/answers.shtml

10   This is a pseudonym.

11   On 28 May 1967 a federal referendum gave the Commonwealth constitutional powers to legislate on Aboriginal matters by amending Section 51 (xxvi) of the Constitution (which gave the States such powers). The 1967 referendum authorized the deletion of Section 127 of the Australian Constitution so that Indigenous people could be counted in the census (Mickler 1998, 121).

12   The 'Aboriginal Embassy' was widely demonized and eventually 'violently removed by police under a new law…introduced by the Liberal government' (Mickler 1998, 139).

13   Real-estate agents selling property in the area do, however, speak of this potential with prospective home-buyers.

14   ATSIC was disestablished in late 2004, and has not been replaced.

15   I live within earshot of the site but somehow slept through the 'Sickening Riots' that 'Rocked Sydney' (*SMH*, 16 February 2004, 1), and resulted in '50 Police Injured during Redfern Riot' (*ABC News online*, 16 February 2004, 1). The images of fires, riot gear, water cannons, violence and injuries that were splattered across television screens and newspapers around the world prompted concerned phone calls and emails, from here, in Australia, and abroad (see note 4).

16   This was quite a frightening research interview because the interviewee was very aggressive, and was 'known' to have connections with racist organizations. I was very careful with how I couched my questions.

17 Aboriginal peoples from the East of Australia are collectively known as Koori.

18 'Cop it sweet' is a play on words, the colloquialisms of 'cops' (police), and 'copping it sweet' ('taking it on the chin' or 'grin and bear it').

19 Between 300 and 400 Aboriginal people fought in World War I and were included in the Australian 'digger legend' (of the bravery and mateship of soldiers). In World War II special units were created for Aboriginal people and Torres Strait Islanders (Bulbeck 1993, 224).

20 'Chippo' is the locally used shortening of Chippendale. *Chippo Politics* was considered to be 'left wing' because of its origins and associations with the Australian Labor Party. It grew out of 'Chippo Politics Forums', which were regular events held at a local pub where guest speakers, of varying political persuasions, would discuss current affairs. *Chippo Politics* gradually changed and is now the *South Sydney Herald*.

21 This was relocated in early 2005 to the ground floor of the TNT building, which is adjacent to Redfern Railway Station.

22 Who was replaced in November 2000.

23 This draws on the use of the term 'race' in reference to Aboriginal people and issues, for example, the 'race debates' of the Hanson–Howard era in Australian federal politics in the late 1990s.

24 I acknowledge John Fielder's (1995) note, in reference to his earlier paper (1991, Gooninup/The Old swan Brewery), that the problem with the 'dirt and danger' analysis is a tendency to universalize and gloss over specific social historical complexities and contradictions. He corrects this by using Stratton's (1990) 'shifts in epistemes of representation . . . and identifies tendencies rather than direct elements' (p. 103, 104). The 'dirt and danger' analysis presented here is an example of 'Aboriginalism' (Attwood and Arnold 1992, 3): 'a specific universalism that racializes the Aboriginal social body, making "Aborigines" out of the Indigenous Australian population'.

25 With the most recent threat by the NSW State Government's 'Redfern Waterloo Authority' his position – to save The Block – is very clear.

26 There are divisions in all communities and it is a mistake to think that Aboriginal Australians are unified about the existence of The Block, or any other issue for that matter.

27 The common sight around The Block is not people 'going wild' as heroin tends to put people to sleep. The 'wildness' may be the association with lawlessness. This interviewee had heard stories of 'bag snatches' and mentioned such fears during the research interview.

28 All currency is in Australian dollars unless otherwise stated.

29 Tasmania is the island State to the South of the Australian mainland. There are Tasmanian Aboriginal people who are descendants of the early 'Tasmanians' (both Aboriginal and non-Aboriginal).

30 ABC TV News, 18 October 2006.

31 In early 2000 a mattress-camp was set up under the eaves of the Aboriginal Housing Company's office building on the corner of Eveleigh and Lawson Streets. It was 'moved on' and the area was temporarily enclosed with storm-wire fencing to stop camping in that spot. In October 2000 a row of terrace houses were demolished and tents became a more common sight.

32 An anonymous informant was a witness to the deliberate vandalism including setting squatted houses on fire. Mostly, however, the fires are campfires around which people congregate in the evenings.

3

'The Good Old Days'

This past doesn't just endure: it displays itself against the tawdry present which it also actively indicts (Wright 1997, 106).

Heritage has gained popularity in many post-industrial inner cities around the world. However, this interest in things from the past is often more than simply a desire to preserve nice 'old stuff'. This chapter tracks the rise of contemporary heritage impulses and the roles such impulses have had, and continue to play in the consolidation of whiteness. It does this by tracing one of the two key types of urban transformation that have reproduced a historical and geographical story of racialization in a transforming residential city. The rebirth of inner Sydney as desirable heritage real estate, the rise in heritage appreciation and protectionism, have proved to be rich fields for the study of the complexity of processes of whiteness-in-action. The second part of this chapter builds on this reading of heritage selectivity, appreciation and preservation by tracing the rise of localized resident activisms that have become integral to the protection of what are widely understood to be simple entitlements to urban spaces, places and at times, buildings. Such entitlements, and the associated mobilizations to protect them, produce what I believe are processes of neo-colonial whiteness. As with the escalation of interest in things historic in Australia – of old Englishness in particular – the examples of resident activism presented here have produced very material outcomes that, on first glance, may appear to be 'common sense', or expected. On closer inspection, such activisms have worked to reinforce and consolidate entitlements that have occurred at the expense of Indigeneity in Australia. This belief in entitlement began at the moment of invasion, with the seizure of lands in 1788, and has translated to struggles for urban space, and places, in the recent past, and continues to manifest in the present.

In the following, I will demonstrate some of the ways that a politics of whiteness has operated within constructions of urban heritage, and associated resident activisms. Some activities reveal that the participants of these activisms have been simply unable to be inclusive of anything other than (neo-)colonial heritage(s) within their charter of protectionism, while others have indulged in more directly malevolent anti-Block activities. These capacities of whiteness include the denial of, or indifference to the histories, and presence, of others. As Herzfeld astutely observed (in Rapport 1995, 645–6):

> Indifference is socially created through the selective rhetorical deployment of a kin-based discrimination between insiders and outsiders (those 'out of place' then being treated 'like dirt').

Denial of and indifference to heritages 'other' than those associated with the still dominant, formally recorded histories of the past[1] persist in inner Sydney. Non-Aboriginal residents continually mobilize to defend what they believe are *their* entitlements to urban space and built heritage. They do this by paying particular attention to the socialities and manufactured memories of 'kinship'. For example, there are members of resident action groups who mobilize to protect built heritage, but there also those who are the non-participating recipients of the benefits of such activism. Just like non-unionized colleagues, who have benefited from trade union actions and the contributions of membership dues – be they the benefits of better wages, working conditions or (of particular relevance in recent times) job security, non-financial participants have enjoyed the rewards of the identity 'kinships' of job descriptions. In inner Sydney, such kinships have enabled the collective valorization of histories that have been deliberately or actively indifferent to all others, to the benefit of those who are not identified as the racialized (Aboriginal, or immigrant) 'others', and identify with common understandings of 'heritage'.

As the neighbourhoods that surround The Block have transformed into desirable space, identifications of 'heritage' have become central to this newfound status. The mostly non-Aboriginal gentrifiers are drawn to this previously avoided part of the city for heritage housing that is still comparatively reasonably priced and consequently marketed for its investment 'potential'. As heritage imaginaries become part of gentrification's cultural capital, they are often formalized. With the early rumblings of gentrification, and its stock of mostly un-improved Victorian terraced houses, the Redfern area gained heritage significance in 1996, when it acquired a 'heritage and conservation' area[2] designation (South Sydney Local Environment Plan, 1996). It was then listed on the Australian Heritage Commission's National Register.[3] In sharp contrast, the Victorian terraced houses on The Block remain very run-down. Some are derelict; many have been razed to the ground. Regardless of popular beliefs and portrayals of its decline, new arrivals to the area find that not only does The Block persist, it also flouts an emergent convention about the sanctity of Victoriana (and its cultural capital). The Block is therefore widely considered destructively *anti*-heritage. In this place, the interdependence of culture and capital is performed, and re-performed through 'heritage'. Culture and capital now work hand-in-hand to consolidate the (neo-)colonial in the gentrifying heritage of the residential city.

Heritage Dreaming

The metropolis of Sydney has experienced a familiar renaissance around its centre. The transformation of the Central Business District (CBD) began in the 1990s to accommodate its bourgeoning global/world city status, and increasing demand for city housing. But it is to the older housing areas around the CBD that I turn to in this chapter. The graceful nineteenth-century Victorian terrace houses that dominate these landscapes are increasingly valorized, treasured, and protected. Following this gentrification trend, former industrial areas began to change as well. Property developers have rebuilt or retrofitted warehouses and old factories with apartments/condominiums. Homeowners have now embraced 'industrial chic' or SoHo (South-of-Houston Street, Manhattan) appeal. As in many other cities around the world, landscapes of restored, or restorable old houses and transformed industrial/commercial buildings are now desirable 'heritage'.

As in other cities around the world, the heritage landscapes of inner Sydney speak of a very specific set of histories. In the case of Australia's oldest city, it is a history of British settlement – and its architectures – that is celebrated. These architectural artefacts have become the desirable remnants, the *post*-colonial heritage of colonial and neo-colonial pasts. In this chapter, I recount the invention of this particular settlement *as* heritage, and identify an associated capacity to forget and exclude other, less palatable aspects of the past that lurk within this celebration of part of a nation's history. I recount some of the motivations behind the designation of 'heritage' and how its embrace expands but only within certain parameters. I also detail the rise of resident activism in inner Sydney and its role in the protection of these specific heritages, and the spaces that they inhabit. To do this, I have drawn on events that have invoked particular resident reactions and the innovative ways that spaces, places and built artefacts have been negotiated and protected through these actions. I argue here that these resident mobilizations have built on a history, and ongoing trajectory, of a specific process of 'whiteness'. Such activities serve in the consolidation of identities that *appear* to be under threat in juxtaposition to the Indigenous 'other'. This chapter documents a way of dealing with the unease that has emerged using a mechanism of escape into fantasies of imagination, of pasts when identities and associated power relations between Indigenous and non-Indigenous peoples were clearer, and more thoroughly fixed. Such times existed before the recent establishment of Aboriginal entitlement to land in the city of Sydney.

This chapter builds on existing considerations of the materiality of cultural processes that are integral to urban transformation. With specific reference to gentrification, David Ley (1986) has long advocated the need to consider the role of aesthetics and taste to better understand the consumption of urban housing. Considerations of culture, identity and subjectivity may have created what Loretta Lees referred to as a 'theoretical logjam' (Lees 2000, 390), that constrained economic and cultural geographies of gentrification – for a while – but it has also helped to clear a theoretical path. By asking questions about taste and consumption, the linkages between culture and capital have come into view. For example, following Bourdieu (1984) the production of 'cultural capital' has been well considered, particularly at the local level of gentrification (Jager 1986, Zukin 1982, 1986, 1995; Jackson 1995). And it is recognized that heritage is a form of cultural capital in the gentrification of old urban areas from Harlem, New York City (USA), to Sao Paulo (Brazil), Kyoto (Japan), and Shanghai (China).[4] Beyond a local scale of analysis, globalizing 'commodity cultures'[5] include an increasingly worldwide appreciation for heritage landscapes and associated gentrification cultures.

The globalization of consumption cultures, from café culture to fashion streets, has been linked to the emergence of 'a global society made possible by…communications and mass media' (Clark 2000, 17). Accompanying wholesale urban redevelopment in many cities are new fields of cultural consumption, part of a globally familiar trend in the way city spaces are (re)made and consumed, around the world (Kearns and Philo 1993). An example is the reinvention of inner-urban housing as 'heritage' (cf Hobsbawm and Ranger 1983), which has certainly occurred in inner Sydney. Another example, and the subject of the next chapter in *Cities of Whiteness*, is the re-imaging of a form of Australian housing that was once far less desirable than detached housing (bungalows). Its rebirth, as *apartments*, has completely altered the appeal of urban high-rise residential living. To make sense of this commodification of certain cultural attributes requires that we understand the 'traffic in things' (Jackson 1999, 105), which includes apartments, or heritage housing. Such commodifications produce meanings and, as Peter Jackson (1999b, 105) has remarked, 'such meanings are...frequently coded in terms of various forms of social difference' or distinction.

Questions of 'difference' from dominant groups has moved beyond the preoccupation with class, and expanded to include questions about 'race' and ethnicity, gender and sexuality (for example, Bondi 1991, on gender, Castells 1983, Lauria and Knopp 1985 on gay gentrification). The emphases then opened to include the production of normative values and spatialities (Lees 1996, Fincher and Jacobs 1998). The addition of a post-colonial perspective has added a distinctive dimension to studies of urban transformation (Jacobs 1996). And, in Australia, such perspectives have brought colonial history into contact with current issues of, for instance, Aboriginal dispossession and poverty.

At the heritage–gentrification nexus, socio-cultural process can work to privilege, or dispossess. In Australia, desires for specific kinds of heritage continue to evolve, and yet the main heritage orbit continues to marginalize migrant and Indigenous heritages. These other heritages have not, however, been simply forgotten. In the examples that follow, I demonstrate how other heritages have been subject to active forgetting; they have been denied through the production of specifically coded understandings of heritage(s) and entitlement that reinforce and consolidate already empowered groups. This is particularly the case for those identifying with – or ascribing to – the privileges of whiteness. In the following, I examine the production of whiteness – how it is imagined and re-imagined. As David Lambert (2001) observed in his study of the liminal status of 'poor whites' in colonial Barbados, whiteness has been made and re-made, and on occasion even *un*-made.

I have divided this chapter into three main sections. The first, 'Performing Sydney Heritage', traces the emergence of heritage impulses in the Sydney context. It considers the valorization of *select* pasts, and objects (built environment and artefacts) associated with these pasts.

The dynamism of heritage production in inner Sydney is then detailed. From Victorian to industrial (façade) landscapes, to retro-chic and heritage referencing in new buildings, specific codes of desirability, taste and specific nostalgias determine what *does* and what *does not* constitute 'heritage'. The analysis includes a discussion of how heritage desires have pushed a cultural politics within the somewhat less traversed terrain of gentrification's post colonial condition.

The second section, 'Heritage Action!' details the rise of resident activism with the gentrification of inner Sydney spaces around the Aboriginal settlement of The Block. Mobilized around protectionism – of people and property – the exclusionary impacts of such activisms are unveiled. With the expansion of heritage impulses, mobilizations of heritage protection and entitlements have also increased. The kinds of heritage protectionism, considered here, have become highly selective. With the rise of 'industrial heritage' appreciation, artefacts of the former (non-Indigenous) 'working classes' have recently appeared on the heritage list. Such memorialization has relied on the commodification of specific artefacts – the façades of old warehouses and factories – for consumption by the 'new middle classes' who invest in their preservation. This section also considers the unexpected fortunes of one specific movement in emancipatory politics through the contributions of Sydney's infamous Green Bans to the contemporary heritage protection movement.

The last and concluding section of this chapter, 'Architectures of Escape 1: Into the Past' considers the preoccupation with colonial and neo-colonial heritages in a post-colonial city. Then the following chapter considers another form of escapism from the unpleasantnesses of now, and here (around Redfern).

Performing Sydney Heritage

> Certain places may be incorporated into sanctioned views of the national heritage while others may be seen as a threat to the national imaginary and are suppressed or obliterated (Jacobs 1996, 35).

The British colonial past looms large in Australian heritage. The push of progress and redevelopment in Sydney, which began in the 1960s, catalysed a growing sense of preservation urgency, by the 1970s.[6] Then, in another burst of (re)development of the predominantly industrial areas of Pyrmont and Ultimo – which was fuelled by the 2000 Olympic Games and increased demands for urban accommodation – the wholesale threat to Sydney's industrial built landscapes was suddenly very real. When fires ravaged several of the remaining large and decoratively historic warehouses earmarked for restoration in Pyrmont, Sydneysiders started to panic. When the proposed development of an apartment block (dubbed 'The Toaster') threatened to obscure a famous vista from the CBD to the iconic Sydney Opera House, anti-development and heritage preservation sentiments escalated. Sydney's heritage became a *cause célèbre*, which garnered the support of celebrities – actors, and other high-profile individuals such as retired politicians and trade union identities – who all joined and promoted various protests. The heritage landscapes that remain are more than precious, for some they are sacred (Taylor 1994). The cultural capital of heritage (Jager 1986) has been realized, through scarcity, and heritage buildings in Sydney have become expensive, exclusive and classy.

Conservation impulses have emerged in response to the mass destruction of modernist urban rebuilding – particularly in Sydney – but the motivations behind heritage impulses vary (Crang 1994). As elsewhere, the widespread enthusiasm for heritage in Australia has enabled the development of a 'cultural heritage movement' and professionals now advise on what *should* be retained, and how to preserve what is left of just over two hundred years of settlement. There are interest groups, such as historical societies, that promote the heritage of old churches, the homes of important historical figures, and so on, while the 'heritage industry' engages in a range of other heritage-related/preservation activities (Hewison 1987, Graham et al. 2000). Meanwhile, real-estate marketeers now 'sell' heritage in marketing campaigns for homes in heritage and conservation areas, such as those that are found around Sydney's CBD.

Historian Ken Taylor (1994) has identified a number of factors that have influenced the emergence of the cultural heritage movement in Australia. These include a reaction to the former lack of interest in history that occurred with the massive urban redevelopments of the 1960s and 1970s which substantially altered the built environments of various cities. For Taylor, a second influence was the resurgence of nationalism, and identity building, in the 1970s. Third, was the development of professional heritage management and public recognition of its potential, which was boosted in 1988 when government monies flowed in support of the Australian Bicentennial. The Australian Heritage Commission (AHC) and the AHC Act of 1975 and the Register of the National Estate and legislation have also provided legitimacy for the protection of heritage. In addition, the reification of heritage tourism, and the rise of heritage awareness and protection that has emerged with gentrification, all point to a thriving heritage industry. Its marketability, or the profitability of nostalgia (Lowenthal 1985) is not lost on big business either, and transnational corporations now occupy expensive historic buildings in downtown Sydney for the somewhat ironic function of selling North American style fast food. Residential property developers have also benefited through re-visioning old industrial areas into heritage-referenced 'warehouse' or New York style loft apartments.

According to Taylor (1994), the current nostalgia for the past in Australia is a search for identity and a reversal of the 'cultural cringe'.[7] He has described heritage as the 'ordinary sacred places ... [that] ... reflect our relationships with places that have meaning because either we, or our ancestors, have connections with them' (Taylor 1994, 27). It is common practice to think of Australia as lacking a recognisable, independent lengthy history upon which to build a national identity (Taylor 1994), regardless of its Indigenous past. 'We' (Australians) are therefore tasked with creating our own (heritage/identity), based on elements of connection that were imported in during the waves of migrations that began a little more than two hundred years ago.

Your heritage or mine?

The designation of what constitutes 'heritage' is not universal. Reflecting on Lowenthal's (1979) contention (quoted in Shaw and Jones 1997) that 'things worth saving need ... [to be] familiar or well loved', Shaw and Jones (1997, 3) responded with: 'today the same place or building can be variously viewed as a homely landmark, a relic of imperial oppression and a tempting commercial opportunity.'

A proposal to redevelop an old brewery site into a large commercial concern in Perth (Western Australia) was highly contested, and provides a useful example of how different preservation impulses not only exist but, at times, conflict. In this example, Indigenous and non-Indigenous 'heritages' clashed (Mickler 1991, Muecke 1992, Jacobs 1996, Shaw and Jones 1997). In one version of 'heritage', an old brewery building was deemed significant and worthy of preservation. In the other version, the site was regarded to be significant, not the building. A spiritual ancestor (the Waugal serpent) of the local Indigenous people resided there – it was a sacred place (Jacobs 1996). The compromise meant that one version of heritage would have to dominate, which it did. The building was maintained because, according to Shaw and Jones (1997), the fate of the site remained ultimately 'in white hands'.

Heritage is more than a reuse of the past. Lowenthal (1985), Jager (1986) and Rosaldo (1989) have all examined the relationship between yearnings for the past, and power. Jager (1986) has considered the operation of class in heritage appreciation. As property developers know all too well, urban conservation can reuse history for the saleable purpose of social distinction. Heritage appreciation, preservation and protection can be bound up in yearnings or nostalgias for times gone by. Such yearnings and nostalgias are often far from neutral, or innocent, as Rosaldo (1989, 108) has also observed:

> 'We' valorize innovation and then yearn for more stable worlds...in any of its versions, imperialist nostalgia uses a pose of 'innocent yearning' both to capture people's imaginations and to conceal its complicity with often brutal domination.

The complexities of race relations and heritage impulses in Sydney are riddled with nostalgias for imperial/colonial pasts. Where yearnings do turn to Indigenous heritage(s), they generally concentrate on those from the *pre*-past, locked away in pre-history and the imaginations of archaeology. Referring to Australian Aboriginal heritage, Taylor (1994, 32) has enthused about 'the Aboriginal wonders of Kakadu...the ordinary places such as Aboriginal tracks commemorating thousands of years of human relationship with the landscape'. This perspective echoes an appreciation for the archaeological status of Indigenous heritage(s) but does little for contemporary Indigeneity. By contrast, the trajectory of British-based colonial European-ness, and it inheritances, has gained a dynamic status, as driving forward into the present. This heritage-dynamic, of Indigeneity as ancient and exotic, with a temporal fluidity for the heritage of the colonizer, has been played out at the 'birthplace' of colonial Australia.

Tony Bennett (1993) has traced the selective preservation and restoration of a popular and well-known historic part of Sydney city known as The Rocks, just near Sydney Harbour Bridge. Since its loving and careful restoration, The Rocks now represents an idealized history of colonization. In this clever rendering, the past was fabricated and cleansed of the marks that bore testimony to other aspects of the history of colonization. 'The glittering façade...functions as an institutional mode of forgetting' (Bennett 1993, 225). The new allegory is the ascent of 'a free, democratic, multicultural citizenry' (Bennett 1993, 227) with Aboriginality referenced only in 'traditional artefacts' that are available for purchased from various retail outlets. The notable lack of an actual Aboriginal presence has served to continually reinforce a commonly held belief that 'European civilization' has prevailed. Wild and natural Indigeneity had been removed; driven out with civilization. The reality of invasion and bloody resistance has been successfully buried beneath the re-written layers.

Back on The Block, in Sydney, the Aboriginal presence does not lie quietly buried – it is undeniably present. Unlike the Perth example, there is no Waugal-like ancestral claim to this place to build a picture of antiquity, and pre-history. In stark contrast, The Block community consists of many kinship groups from around the country.[8] It is a strategic meeting-place for the most dispossessed of such lineages, with representation by members of the 'stolen generations' of Aboriginal peoples. Because of its overt presence, in the present, this form of urban Indigeneity blatantly resists any pre-historic casting. Aboriginal people fought for The Block in the 1970s; it was *won* as part of an emergent 'black politics' in Australia. And its highly significant heritage has gained official recognition. The Australian Heritage Commission announced that they had listed The Block on their National Heritage Register of Australia as a site of Indigenous significance, on 25 October 2000, and broadcast this decision on the television news, that evening. However, the dominant understanding of heritage that surrounds the site refuses to acknowledged this significance. Instead it remains fixed on the built heritage of houses, large and small, factories and warehouses. In what follows, I will argue that this is far from accidental. It is part of a neo-colonial manoeuvre that continues to operate in the ongoing service of inner Sydney whiteness.

Glorious Victoriana and other 'old stuff'

'[the city is] a poem...which unfolds...and it is this unfolding that ultimately the semiology of the city should try to grasp'. (Barthes 1986)

Cities can be 'read', as systems of communications, which can tell us 'who has power and how it is wielded' (Short 1996, 390). Urban theorists and researchers now recognize the symbolic and representational realms (figurative and discursive) in the constitution and mediation of social and material processes (Jacobs 1993). Meaning is 'hidden' within, for example, architecture, or artefacts (Crilley 1993). These hidden meanings can be also be 'read' and variously interpreted, and re-interpreted (Barnes and Duncan 1992).

Although Sydney's gentrification is well documented, the analyses have tended to follow a political economy approach (Keating 1991, Horvath and Engels 1985, Engels 1994, 1999, Bonyhady 1995). Stories of the revitalization of formerly grand suburbs, such as Paddington (and Woollahra), which have become *the* benchmark for Sydney gentrification (Cameron and Craig 1985), are the most common. Such accounts (except Powell 1967) have rarely considered the rise of heritage appreciation with gentrification.

As inner cities become increasingly attractive, heritage becomes valuable. Bourdieu's (1984) notion of cultural or symbolic capital considered the 'social uses of art and culture and the way that "tastes" function as markers of "class"' (Bourdieu in Jackson 1991, 220). As gentrification cycles mature, the cultural capital of heritage becomes increasingly attractive. Heritage then demands higher premiums. As has happened elsewhere, those seeking cheap housing in inner Sydney – for the purposes of adding 'sweat equity', or for rental, have been increasingly sidelined.[9] Neil Smith (1987) identified 'second stage' or 'yuppie' gentrifiers as those who (can and will) pay a premium for heritage. These gentrifiers are distinct from 'first stage' gentrifiers, who buy into less desirable inner-city areas at low cost when un-renovated or 'poorly' renovated houses are plentiful. Heritage and preservation issues may be of interest at that stage but ultimately, first-stage buyers (re)produce heritage capital simply through their efforts to save and restore the built fabric of decaying buildings. First stage gentrifiers serve to provide legitimacy for living in such housing, and establish the heritage pedigrees that second stage gentrifiers then seek. For classic examples, think of Chelsea or Islington in London or, in a slightly different version, SoHo in New York City.

In Sydney, the neighbourhoods adjacent to The Block were largely impervious to gentrification[10] until the mid- to late 1990s when the stocks of 'unimproved' (unrenovated) Victorian terrace houses diminished elsewhere. Although the tardiness of Darlington's gentrification has been attributed to a range of factors, such as the mix of (long-defunct) industry with housing, the presence of public/social (state-owned) housing and the lack of retail services, a planning survey revealed something else. According to Cameron and Craig (1985, 29), the significance of 'the presence of ethnic and racial groups and...associated physical violence and social tension' was also hindering gentrification. At the time of the survey, 16 per cent of terrace housing in the Darlington/Redfern area remained unimproved, while in Paddington, only 7 per cent remained so (Cameron and Craig, 1985, 25). The scarcity of 'unimproved' housing stock indicated that Paddington was in an advanced stage of the gentrification. At this sate, unimproved houses attracted premiums that only those with the economic power to put economic necessity at a distance could afford (Bourdieu 1984). In Sydney, 'original condition' is a selling point for heritage housing. It implies that most of the decorative heritage features are intact. 'Victorian features' is a phrase also favoured by real-estate agents. Of course, there are many houses renovated in the 'wrong' style, and those with a 'Mediterranean finish' (Howe 1994, 155) tend to be avoided by heritage-seeking home-buyers.

Like the occupation on The Block, Sydney's 'migrant' occupation of the inner city has been relegated to the domain of anti-heritage (Armstrong 1994, Lozanovksa 1994). Post-war migrants 'rescued the reputation of the terrace house as a place to live as well as restored its fabric' (Howe 1994, 155), just like first-stage gentrifiers'. But around inner Sydney, as elsewhere, real-estate agents continue to associate migrant additions and renovations with 'tastelessness'. The replacement of rotting timber windows with aluminium, and collapsing timber floors with concrete is now generally thought of as vandalism. Houses with such features are said to be 'wogged' (Research Interview, 4 March 1998). Heritage, on the other hand, is associated with the restoration of timber windows and polished wooden floors. The modifications that migrants from Southern Europe made to ageing housing stock, to make them more habitable and culturally appropriate (Lozanovksa 1994), now constitute a deficit in heritage capital. New owners usually remove this migrant layer of history. Intact Victorian architectural features, without the hindrance of other layers of history, have become a priority for those who can afford them.

From Grand Victorian to working-class heritage

By 1999, house prices in inner Sydney had risen substantially[11] and premiums for terrace houses in Darlington/Redfern, reached a new high, regardless of their condition. By this stage, any renovation (other than immaculate heritage restoration) of houses had become a detractor to sales, and vendors often found that they had over-capitalized.[12] By late 1999, another trend started – this was the conversion of small[13] terrace houses into 'urban living spaces'. In Darlington, many gutted their smaller houses, leaving only the façades of these highly capitalized properties, often retained only to meet heritage regulations.[14] This trend, where the sky seemed to be the limit on interiors, indicates that investors were confident enough to outlay large sums on small land parcels, regardless of the area's insalubrious reputation. Additionally, retaining the heritage features of the interiors of these houses, formerly inhabited by the working classes, was not nearly as important as what they could be become (Figure 3.1).

In Darlington, the recent rise of heritage significance of the smaller 'working-class' houses, or 'workers' cottages', can be aligned with scarcity in larger housing stock. Memories of the diminishing working classes are fading, but some have promoted them as part of the folklore (and therefore, cultural capital) of the inner city. For example, the (retired) high-profile (Labor) Mayor of South Sydney City Council proudly promoted *his* 'working-class' heritage. Local newsletters, such as *Chippo Politics*[15] and the Council's *South Sydney Inner-city News*, have also carried on this tradition. As the built heritage is consolidated nostalgias 'write in' the histories of those that were less important (Crang 1994). For one resident, however, the smaller terrace houses

Figure 3.1   Restored Workers' Cottages in Sydney

were 'an ugly reminder of dirt-floor basic housing...that was...not very pleasant' (Interview, 25 April, 1998). When prettied up, these 'ugly reminders' become transformed into postmodern extravaganzas. Within the pastiche of their new presentation, their heritage value is clear – it lies in their external shell and occasional quaint attribute, such as a salvaged 'dunny' (brick shed housing original toilet) out the back, which makes a handy garden shed.

Industrial façadism, retro-chic and new-build heritage

In another turn in the heritage trajectory, the industrial landscape of Chippendale, which borders The Block to the North, was described as a 'dead space' in 1997 (North and Christie 1997). It was then reborn, in 2000, as Sydney's own version of New York's SoHo (Shaw 2006). Astute property developers filled old warehouses and factories with loft apartments/condominiums and this formerly disregarded area is now an important industrial-heritage landscape.

Similar to some of the smaller workers' cottages, in this version of heritage a remnant, such as the shell of the old building, is kept. In this façadism compromise (between retaining a whole heritage building and its complete replacement), which demonstrates that a little bit of heritage is better than none, it is the *purpose* of the factory – its business – that is celebrated.[16] This form of industrial heritage fetishizes objects connected to the factory or warehouse, such as the bicycles or pianos once made within. Industrial architectures, and remnants of some of the larger industrial aspects of the interiors, such as large wooden or steel beams, and the occasional block-and-tackle removed from its original position to take pride of place in an apartment lobby, are part of the promotional aspects used to sell the apartments that have been fitted within these structures. The promotion of nostalgias for objects helps to partition the past. The heritage of the 'battlers' that occupied the interior as workers is not recalled. The class (and often 'race') relations, the working conditions then and displacements now, are easily truncated and buried in history (Jenks 1981). For Rosaldo (1989), such partitioning has the capacity to allow, or even participate in the repetition of forgotten violences. For some of the older folk, who lament the loss of places such as 'Miss Muffett's jam-making factory' (Interview, 27 April, 1998), those 'dirt floor' days were actually *the good old days* that have been quickly forgotten.

Meanwhile, the heritage recognition and consumption circuit (Jager 1986) continues to expand but not in the direction of The Block. From grand Victoriana, to the inclusion of small Victorian workers' cottages, and more recently built factories, two other players, 'retro-chic' and 'new-build' faux-heritage, have gained heritage status in inner Sydney. Retro chic includes an appreciation of old things, objects that, in Patrick Wright's (1997, 104) words, are 'not-quite-antiques' but nonetheless collectable. Definitions of gentrification heritage have expanded to include, for example, Art Deco architecture, and furnishings, and the more recent addition of 'retro-modern' style from the 1950s and 1960s (Collins 1995). Newtown, which adjoins Darlington to the south, has become *the* retro-modern and *objects d'art* centre of Sydney, and is comparable with the concentration of more standard antique shops in Paddington and Woollahra that specialize in 'top end' antiques. The emerging preference for retro-modern furnishings and items, particularly from the 1950s and 1960s, which are cheaper than antiques (defined as at least 100 years old), signifies a diversity and distinctiveness of tastes (for objects of desire) that exist within gentrification cycles.

Amid the heritage built environments of inner Sydney, heritage-referencing or faux-heritage, has also gained currency. Although not heritage *per se*, these forms of 'neo-archaism' (Jager 1986, 88) have found expression in new developments, particularly with apartment conversions and new terraces. Along with the very old parts, heritage references pepper the new brickwork, the cobbled laneways, and the stone gutters. There is a preference for faux Victoriana, as heritage-referenced architecture is generally preferred to non-heritage-referenced building designs.[17]

As the meaning of 'heritage' continues to expand, so too do desires for built heritage, and preservation. These increasingly mainstreamed desires prioritize the symbols of select pasts, of terrace houses and old industrial façades, over contemporary expressions of human diversity. In inner Sydney, resident action groups mobilize to fight developments that are regarded to be 'tasteless' (non-heritage referenced) and/or dedicated to housing *others*, such as students from abroad. Such NIMBY preoccupations, which have tended to be dominated by self-interested protectionism (Dear 1992), have assisted in the disengagement from, and concern about human diversity. Yet, it is 'diversity' that makes up part of the lexicon of expanded heritage designations. One of the roles of 'the community' is to continually monitor for threats, and engage in protection of old architectural diversity, as required. At the same time, desires to protect human diversity (such as the memory of those may have toiled behind the protected façade, the migrant 'others', or the Aboriginal community around the corner) are sidelined by concerns about *who* will occupy new developments. For example, students from overseas are clearly not wanted (Public Meeting with South Sydney Council, Saturday, 14 June 1997). The purpose-built student housing development, by UTS (discussed in previous chapter) became the benchmark of undesirable 'tastelessness' for the area. Heritage and taste intertwine, and percolate through the discourses of protest against such developments, which have been labelled as *anti* heritage. Darrel Crilley (1993) identified that property developers promote diversity to enhance the appeal *of* developments, rather than to appeal *to* a diverse market. The appearance of diversity, as expressed through an expanding range of 'heritage' architectures, has the capacity to conceal homogeneity. 'Diversity', in this case, is simply another consumable attribute for affluent tastes and rather than appealing to a range of types of people, only those with the necessary attributes (such as cash, class and/or the right ethnicity) have membership in such a niche market.

The preoccupation with protecting symbols from the (neo-)colonial past(s), or allowing only those developments that are deemed tasteful, has enacted an architecture of denial of human diversity, as the expanding orbit of heritage designations continues on its trajectory of inclusion, and exclusion. The deeply embedded desires to preserve and protect (colonial remnants) have become an escape from everyday realities, which include overt Aboriginal poverty and dispossession just around the corner from a tasteful 'heritage' built environments. The notion of heritage, as it is popularly conceived, and as governments have legislated generally,[18] exhibits a certain consistency. Heritage remains commonly associated with old buildings, and objects.

The unspoken heritage story tells the history of colonization, of encounters between Aboriginal and non-Aboriginal people. The heritage of The Block, and its struggle for Aboriginal civil rights, remains actively excluded from heritage understandings regardless of its formal heritage status. There is no commemorative plaque, or acknowledgement that this site has finally been recognized as a site of *cultural* heritage. Heritage has become part of the unspoken definition of 'community', of belonging. For the area around The Block, heritage remains architectural/artefactual, and where people *are* considered, it is the elegant lives of those who could afford High Victoriana and, in a more recent working of heritage, it is a partial legacy of what is remembered as 'white' working-class ('battler') housing, and the industries that flourished. These yearnings for 'more of the same' (pasts) are sometimes whispered within such imaginings; they are collectively remembered (Boyer 1998).

While the heritages of Aboriginality may be acknowledged, such understandings remain tied to non-urban, pre-colonial times. The politics of the 'Black Capital' (The Block), of the unification of disparately dispossessed Aboriginal peoples – who have united under the now well-recognized Aboriginal flag (conceived in 1971) – is continually written out of the evolving urban heritage imaginary. The non-Aboriginal, non-migrant gentrifier imaginings of heritage are thereby excused from engagement with the urban settlements of others.

Activating Heritage

Inner Sydney residents (and others), mount regular campaigns to preserve heritage, particularly when old buildings or urban spaces come under threat. In this section, I elaborate on the theme of selective heritage(s) by tracing the contributions of localized resident activism to projects of whiteness. The following gives examples of the valorization of non-Aboriginal urban landscapes, through resident activism, and the active devaluing of the Aboriginal settlement of The Block. Such activisms have produced material outcomes that have had either direct or indirect impacts on The Block.

As noted in the previous chapter, many non-Aboriginal residents share a sense of endangerment because of the Aboriginal presence. The emergence of associated activisms follows Sidney Plotkin's (1990, 219) idea that in struggles over community land use, 'it is necessary to see one meaning of community as defensiveness and exclusion, as a sense of beleaguered membership in an endangered enclave'. In the following, I detail several specific incidents that reveal the contributions of localized resident activism to broader structures of inequality between Indigenous and non-Indigenous peoples. These activisms have relied on a sense of endangerment to person, to property (and property values) and, most significantly, to cherished heritage(s) and associated entitlements to urban spaces of whiteness. The history of embattlement, the protection of 'turf', and the retention of some of the façades of old workplaces, has helped to retain idealized images of the past. Within the evolving, and multiply faceted understanding of heritage there is also a telling ambivalence about some aspects of heritage. For example, 'industrial' heritage exists largely in the façades of old buildings. Far more important than where workers toiled for a living is the role of the industry. The 'battlers' may have played a part in the preservation of housing and territory – they did fight to save the heritage housing and the place that gentrifiers increasingly occupy, but their working lives are less memorable. The fighting battlers are remembered, however, and their capacity to wage a battle is their legacy, which is recalled as required, in the cultural practices of city living.

Urban social movements, resident activism and enclave consciousness

Localized resident activism began to re-shape urban processes and form, in cities around the world, in the 1960s. Along with other 'movements', urban social movements began to challenge the policies or actions of the state (Castells 1977). According to Fincher (1987), early urban social movements were often progressive agents of change. By the early 1980s, however, commentators were noting that localized resident activisms were often motivated by fear, and the need for 'turf protection' (Dear and Taylor 1982). Local resident activisms of this kind have urged a shift in the previous assumption that resident action was largely a politics of class emancipation (Castells 1977, 1983).

In their study of resident action in Sydney, Costello and Dunn (1994) identified that resident activism is either 'diversionary' or 'critically empowering', at a local level. Diversionary activisms are those that can divert or stop a proposal, such as a development proposal. Diversionary activisms are often initiated with land use disputes. Activisms that critically empower, on the other hand, include political movements such as the green movement, women's movement, peace movement, and gay and lesbian liberation struggles. These movements mobilize at a *grass-roots* level. The political ideologies of these latter movements are usually broader-based and distinct from 'protection of turf' motivations (Costello and Dunn 1994).

Dear and Takahashi (1997, 87) categorized Not In My Back Yard-ism, or 'NIMBYism', as either 'reactive . . . to protect existing amenities and distributions of resources', or 'empowering . . . to correct past inequalities'. Although these are somewhat different categorizations of resident activism to those of Costello and Dunn (1994), there is agreement about the politics of urban social movements. Those activities organized around ideologies of localized turf protection, which actively privilege local issues over challenges to larger political structures or dominant ideologies, are likely to be in the grip of the NIMBY syndrome. NIMBY activities tend to be dominated by self-interested protectionism.

It is now clear that some local activisms mobilize for conservative and exclusionary purposes (Plotkin 1990). We recognize the impact that some urban activism has on minority groups, where race and class intersect (Dear and Takahashi 1997), and where endangerment by 'something other' is resisted (Dalby and Mackenzie 1997, 101). The following details various urban social movements and resident activisms in inner Sydney. The examples that follow had direct, or indirect, impacts on the presence and continuation of the Indigenous settlement of The Block.

Enclave consciousness and NIMBY activism

In the late 1960s and early 1970s, two monumental events occurred in inner Sydney. One was an announcement that the University of Sydney was expanding, and the other was the rise of an Aboriginal 'black politics'. Both events sparked resident activism. In response to the announcement by the University, housing in Darlington was threatened, and residents mobilized to save their suburb. The second set of resident activisms were in response to the mobilization of land rights for an Indigenous place in the city of Sydney, and the proposal for an Aboriginal settlement, which would become known as The Block. This section introduces how an emancipatory politics, and the associated struggle for Aboriginal land rights in Sydney, elicited what has become an ongoing and complex movement of racialized resistance.

In the early 1970s, an announcement that heralded the formalization of the Aboriginal presence in the city of Sydney, through the purchase of land and houses near Redfern Railway Station, sparked an immediate response locally. The 'South Sydney Resident's Protection Movement' (SSRPM) formed to fight the 'festering sore' of Aboriginal settlement (Anderson, 1993b, 328). The SSRPM argued that the 'ghettoization' of Aboriginal people would lead to violence and hatred. One particularly feisty group of non-Indigenous residents declared that 'a human zoo should not be allowed in this area' (quoted in Anderson 1993b, 328). The most racialized 'other' was a frightening, uncivilized and, for some, quite beastly entity.[19] The decision to fight the proposal to formalize the Aboriginal settlement was supported by prominent Labor politicians from the local council to the federal government. There was, at that time, a longstanding belief that that Aboriginal people should merge into mainstream (that is, predominantly non-Aboriginal) society.

Sidney Plotkin (1990) has identified the exclusionary politics of 'enclave consciousness' (Plotkin 1990, 223). In the context of this book, the two important emancipatory social movements, of preserving 'working-class' housing (from subsumption by the University) in one case, and the recognition of Aboriginal land rights in the other, were both grassroots movements that were grounded in broader emancipatory politics. On the other hand, the actions of the SSRPM, and more recent local activisms, that have been overtly NIMBY in character, have been motivated by desires to protect 'turf', and heritage environments. According to Plotkin (1990, 223), such 'enclave consciousness' has within it, a sense of besiegement:

> Enclave consciousness sees its primary political conflicts with the outside society, not within…this way of thinking can end up justifying the repression of minorities…while supporting the social power of leading economic interests.

This state of 'consciousness' can be observed in both the battle against the expansion of the University, and the battle against the formation of The Block. In the first battle, the University had become the tyrannical landlord, and land-grabbing overlord. In the second, the Federal Government was the authority that threatened to usher an unwanted Aboriginal settlement into the same embattled neighbourhood. In both battles, unification against the large and external enemies occurred along ('working') class lines. Of course, the main distinction between the two battles is also identifiable. The fight against the Federal Government, and others involved in the development of The Block, resulted in a seemingly justifiable repression of a minority, as identified by Plotkin (1990) in the aforementioned quote. The 'emancipatory' politics were therefore highly selective, and racialized. At that time Aboriginal people were simply too *alien* to be considered as part of the 'working classes', regardless of their employment status. The bigger picture, of Aboriginal dispossession, unemployment, poverty and homelessness was external to the logics of class. Instead, Indigenous peoples were seen to be part of a particularly brutal and frightening form of 'nature'.

In accordance with Plotkin's (1990, 223) definition of 'enclave consciousness' (quoted above), a representative of 'leading economic interests' was also involved during the anti-Block battle. The Block site had been bought by a developer in the same way that the University had bought up swathes of Darlington in the previous territory battle, house by house. Unlike the university context, however, the trade union movement imposed a development ban. Ultimately, the developer had no choice but to sell the site that would become The Block (Anderson 1993b). In the meantime, the localized politics of class emancipation, which had been exhibited in the earlier battle with the University, had slipped into a state of exclusionary enclave consciousness. For the Darlington residents the less than desirable interests of the property developer, which could eventually threaten the poorest of the working classes through spiralling rents, were the lesser of the two evils before them. Ultimately, the threat of the interests of the property developer was not *as* undesirable as the prospect of Aboriginal settlement. In accordance with this, the local council, the now disbanded South Sydney City Council, was strategic. When it received the development proposal for The Block site, it would only approve development of single-family units. Such a stipulation meant that the requirements of the Aboriginal community for extended-family housing would prove to be a stumbling block in the planning process (Anderson 1993a). It seemed that the wider expression of emancipatory politics had succumbed to localized and heavily racialized enclave consciousness.

Events such as these are politically complex. The anti-Block response was a result of a range of reactions to various political machinations occurring at the time. Issues of class emancipation combined with an emergent, and localized, political conservatism that was born out of a sense of betrayal and consequent besiegement, which probably fed into the fears surrounding the unknown 'other', particularly after the incursions by one of the most elite institutions in the country, at that time. It seemed that that an already impoverished and disempowered neighbourhood would simply have to bear such burdens. No *other* urban area was targeted in these ways; no other residents in Sydney would have wanted or accepted the conversion of part of their neighbourhood into Aboriginal land, and would have held a stronger fighting position than the impoverished and disempowered residents of Redfern and Darlington. The legacy for inner Sydney is that these events helped to establish a specifically racialized local politics that continue to this day. The ugly, and now unacceptable representations of Aboriginal animality, and 'human zoos', have been replaced with stereotypes of heroin-addiction and criminality. This newer typecasting has become the basis for ongoing anti-Block hyperbole and particularly malevolent resident activisms. More generally, resident activism in inner Sydney has become a recognisable part of the urban planning process with what seems like a relentless march of apartment development.

Anti-development politics often resemble a form of NIMBYism. It is, however, clear that anti-Block activisms in the 1970s were based on assumptions about *who* would be moving in, rather than the redevelopment of the site. Meanwhile, during the 1970s, another politics – a broader anti-development mobilization – had begun around Sydney, which would also have implications for the development of The Block site. This movement became known as the 'Green Bans' and these bans worked to protect 'working-class' housing from rapid urban re-development. Based on emancipatory ideals, the Green Bans were very distinct from anti-Block NIMBY politics and ultimately worked in favour of the Aboriginal community of inner Sydney.

From Green Bans to heritage protection

The anti-development movement of the 1970s, known as the 'Green Bans' (Jakubowicz 1984) protected homes and green space from a largely unfettered development boom. The politics of the Green Bans – trade-union-imposed work bans placed on specific urban developments – were particularly protective of housing in the old inner-city suburbs of Sydney which were, at that time, where the poorest lived.

The issue of housing low-income earners in the inner city has largely given way to larger economic forces. The 'working classes' have tended to move to the 'affordable' (outer) suburbs. The emancipatory politics of the Green Bans movement therefore had another, somewhat unintended consequence. Because a lot of 'old' housing stock was saved from demolition, a conservation ideology for Sydney's built heritage was ignited. This 'collective urban social movement' involved a diversity of groups involved in the protection of housing and open space. Although having a primary concern with emancipatory politics – the protection of the homes of the working classes from large-scale development, including freeway construction – the Green Bans did pave the way for the expression of the heritage conservation movement, which has – somewhat ironically – flourished with gentrification. Gentrifiers have turned the strategies of 'people power', and the techniques of protest, to the task of protecting heritage architecture. The built fabric of terrace houses, that was once retained to house low-income earners, the 'working classes' and others, is now protected for very different reasons.

The rise of protectionism in inner Sydney

One particular development ban, a quasi- or proto- 'Green Ban' raised more than a few eyebrows. In a gesture of emancipation and recognition of Aboriginal land rights (Anderson 1993a & b), a trade union development ban was applied to the area that became The Block. In an interesting shift in recognition, this ban assisted some of the most disempowered people in Australia who, in this case, were Aboriginal. For the incumbent non-Aboriginal working classes of Redfern and Darlington, the imposition of this development ban was totally unexpected. It was another act of treachery. Such betrayals, first by the Labor Federal Government, and then by the trade union movement (who were behind the Green Bans) laid the foundations for a location-specific sense of siege and enclave consciousness. The following section documents how, from these earlier events, a particularly racialized form of enclave consciousness has emerged, and is maintained. Such a consciousness has evolved, and can be traced back to the advent of seemingly innocuous initiatives, such as Neighbourhood Watch. These helped to establish a range of responses to the existence of The Block, from a system of informal surveillance through to various examples of racist extremism that have emerged from time to time, when entitlements to urban space are deemed to need protection, or reinforcement. Such activities continue, and serve to further agitate and exacerbate anti-Aboriginal sentiments, and ever more creative activisms.

Neighbourhood Watch

The popular anti-crime initiative, Neighbourhood Watch (NW), was attempted in the Darlington/Redfern and Chippendale area, in the mid-1980s. The NW concept, based on 'community control' with local police support, used simple surveillance techniques, such as 'keeping an eye out' to help protect property from theft and damage. There are no records of the Darlington branch of NW, which would normally be kept at the local supporting police station (in this case, Redfern Police Station) but, according to several long-serving police officers (interviewed in 1997), it lasted from 1985 until 1990. There are at least two versions of why NW folded in Darlington. In one version, NW gradually ceased due to waning interest:

> I used to be a coordinator for here... well we kept the Neighbourhood Watch going. The only time you'd get a crowd there was when they were robbed... then it just phased out... what was the good of keeping it up if few people [were] here? (Interview, 27 April, 1998)

Others were less 'neutral' in their response to the demise of NW in Darlington:

> What happened was the police tried to start it and it was taken up [by local residents] to make it look as though every bit of crime that happens in the area is because of the Aboriginal community. It wasn't because the police were trying to do that; it was just there were some people in the area who appropriated it. (Interview, 12 February 1998)

The loss of NW in the Darlington/Redfern area was lamented by some and applauded by others. Accounts of NW programs have provided various diagnoses of the demise of local initiatives, which included changing gender roles and the consequent reduction in networks of home-based women (Hillier 1996, 101). By the late 1980s, racism in policing in the Redfern area had become the focus of an inquiry commissioned by the National Inquiry into Racist Violence, Human Rights and Equal Opportunity (Cunneen 1990). The demise of NW was also linked to this issue:

> There were some really big, very nasty police raids down on The Block and that involved a lot of meetings to develop what now are considered...normal things in terms of Aboriginal police liaison activity [such as regular meetings between police and Aboriginal Elders], which people just didn't have at that stage. (Interview, 12 February 1998)

It would appear that the 'winding down' of the Darlington branch of NW was due to a range of factors including improvements in policing, through new police liaison initiatives. Regardless of how it wound down, the legacy of NW was that surveillance was established; NW was simply a formalized version of an existing culture of surveillance. Local residents had long engaged in 'self-surveillance' – they would watch out for themselves, and each other. This form of 'community' surveillance continues to this day to protect the non-Aboriginal community people, and property. An extension of it occurred in the late 1990s, when a group of residents in Chippendale engaged the services of private security guards to patrol their streets.[20] This professional surveillance was meant to show the local police that residents were serious about surveillance and wanted it to become more formalized through an increase in 'beat patrols' (Interview, 17 August 1998) as many residents felt that the area was seriously under-policed (former Local Area Commander, Pers. Comm., 4 August 1998). The police, non-Aboriginal residents, and business people all watch anyone, and everything, associated with The Block and this occurs in informal and informal ways.

Grassroots activism or reactionism?

In addition to surveillance, specific tactics have been mobilized through resident activism in response to drugs and crime in Redfern and Darlington. The following traces the deployment of two particularly telling tactics. The first served to denote The Block as the completely unacceptable perpetrator of drug-related crime. The second tactic involved the rise of resident activism, which targeted – and thwarted – plans for a government facility for the treatment of drug and alcohol related heath issues in the Redfern area. This second example has served to establish a NIMBY culture that has remained strong and triumphant. These examples are linked through the issue of illicit drug use, and the widely held belief about the potential for drug use to spread from The Block (and by dint of location to The Block) into *non*-Aboriginal spaces.

As I have detailed earlier, The Block is reputed to be almost synonymous with drug use. It is also widely regarded as a place for the acquisition of (heroin and other) drugs, and drug-use equipment. Public concerns about drug use, which resulted in a NSW government 'Drug Summit' in 1999 ('Carr calls summit,' *Sun Herald*, 7 February 1999; 'Carr defends needle exchange program,' *Australian Associated Press* [Reuters], 20 May, 1999), combined with concerns about the plan to trial a version of the controversial Wayside Chapel's safe-injecting room (in Kings Cross), in The Block area. Various resident action groups formed in response to the range of drug-related issues, and one particularly active group that I will refer to as the Kommunity Action Kollective (KAK),[21] engaged in a powerful campaign of racialization.

The birth of KAK

KAK formed in the mid-1990s to lobby against a development proposal to build housing for Aboriginal Elders on The Block. KAK's main objection to the proposal was that the development would encourage 'blow ins' who would seek shelter with the Elders (Interview, 13 September 1998). South Sydney City Council records indicated that this development was approved in 1988[22] (letter from Freedom of Information Coordinator, South Sydney City Council, 17 May 1999). In another campaign, KAK tried to have a laneway closed. The plan was to gain permission to partition the lane for purchase by adjoining households. The lane is adjacent to Eveleigh Street (the main street of The Block), and is notorious for drug use. Selling a section to residents with houses backing onto the lane would have privatized it, and thereby removed the space from public use. The authority with the power to make such a decision, South Sydney City Council, had allegedly agreed to the proposed sale of portions of the lane, for 'about 4 or 5 grand [thousand dollars]' per household (Interview, 13 September 1998). However, residents with properties backing on to the lane on the opposite side objected. The Land and Environment Court of NSW halted the sale of the lane. These residents would have lost access to their properties from the rear (the laneway) if the plan had proceeded.

This was a defeat for KAK. Had they succeeded in the privatization of the laneway, the drug 'shooting gallery' problem would have been defused (or at least diffused!). Second, property values would have benefited by the addition of a portion of land to very small blocks. Third, space that had become a 'no-go zone' to everyone but drug users, would have been captured. A resident summed up the situation in this part of Redfern like this:

> We bought here mainly because of price, and also because of expectations of the area being gentrified, and we feel that it will... and also we're prepared, we are young enough to sit out the 'disputes in Beirut' here and in due course we hope to see the problems dealt with and we'll benefit from that economically and also socially (Interview, 13 September 1998).

This resident expressed a preparedness to 'sit out the disputes in Beirut' and wait for the windfall of gentrification. The use of the Beirut metaphor exemplified a common perception about living in a 'war zone'. When the 'war' ends, however, the winners, in this resident's view, will be the astute and patient young (non-Aboriginal) home-buyers.

The absorption of Caroline Lane into private space would have been an astute economic move in a gentrifying area. Although KAK failed to secure the lane, its members continued to campaign. The consistency of KAK to mobilize, and to morph into other guises during other campaigns (detailed below), has enabled a strong anti-Block position to be heard again, and again. Several events occurred in the late 1990s that were also designed to rid the notorious laneway of what had come to be known as the Aboriginal drug problem.

No Medical Facilities in My Back Yard!

The following details a government response to publicized drug use in the aforementioned laneway. The front page of a popular Sunday tabloid newspaper carried a photograph of a 'young boy' being injected with heroin by an adult in the laneway ('It's Wrong', *Sun Herald* 31 January, 1999). Though not stated (and the ethnicities of both the adult and boy in the photo were ambiguous) the notoriety of the laneway, and its proximity to The Block meant an association with Aboriginal people, and specifically, Aboriginal heroin use. The resulting uproar over this incident elicited an immediate response from the then state Minister for Health and Aboriginal Affairs. According to a newspaper article (headlined 'It's Wrong', *Sun Herald*, 31 January 1999), the minister 'quickly ordered a halt to the official needle-exchange program conducted in . . . Lane'. He is quoted as having said 'I am suspending the program [of needle exchange] . . . as of today and we will hold a top-level investigation into the circumstances surrounding this incident [of the assisted heroin injection]'. Journalists fuelled the outrage by describing the needle exchange program as 'a scheme that hands out heroin injection kits to children' (*The Sun Herald*, 31 January 1999). Somewhat ironically, the photograph resulted in the removal of a government sponsored needle-exchange service in an area that seemed to desperately need this facility.

According to Dr Alex Wodak, the director of the drug-and-alcohol service at one of Sydney's largest medical units, St Vincent's Hospital, the needle-exchange facility had been 'an enormous success in cutting the number of drug addicts getting HIV' ('It's Wrong', *Sun Herald*, 31 January, 1999). Then, a week after the furore, it was revealed that the 'child', who had been estimated to be about 12 years of age in the newspaper article, was actually 16 years old. This media feast continued with a series of reports about the criticisms of the needle-exchange *closure* by health authorities ('Experts condemn needle shutdown', *SMH*, 1 February, 1999, 3'; 'Redfern closure wrong', *Canberra Times*, 2 February, 1999, 8). The needle-exchange service had been removed from the laneway[23] because it was deemed to be promoting the dangerous practice of drug use in an Aboriginal drug-use area. A local Aboriginal spokesperson for The Block, Lyall Munro, saw the event somewhat differently, and asked: 'why wasn't the media there when Aboriginal kids were shooting up? . . . They have the same problems as that white kid who was photographed' ('Heroin shot – Aboriginal spokesperson says he smells a rat', *Australian Associated Press*, 31 January 1999).

Interestingly, two KAK members happened to be present on the day the photograph was taken and their comments were also reported: 'It's not a needle exchange, it's a needle give-away' said a resident... who discovered the needles' after an opened box of syringes was found in the place of the removed needle exchange unit. 'I know... [the lane] is a disgrace. We're trying our hardest but it's an uphill battle', another resident said ('Needle give-away', *Daily Telegraph*, 2 February 1999, 5).

This quote was informative for reasons other than what the residents said (about the health service provisions for drug users). The newspaper article used the first names of these two individuals,[24] and the Darlington community was very familiar at that time with both personalities because of their anti-Block doorknock appeals. Both were also members of KAK and its offshoot groups and committees. One of these 'committees' was responsible for a campaign that collected, allegedly, 500 'crime victim' stories around the local area. These stories were used in gaining signatures for a *quasi*-petition (Figure 3.2). The target of the petition was not specified but these 'petitions' were sent off without the knowledge or consent of the residents who had provided the stories, to a high

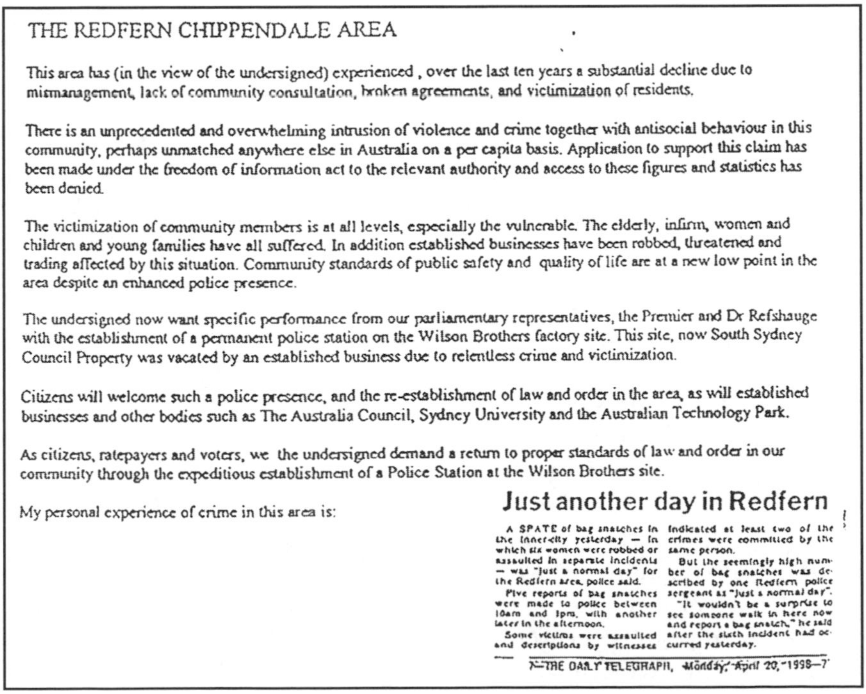

THE REDFERN CHIPPENDALE AREA

This area has (in the view of the undersigned) experienced , over the last ten years a substantial decline due to mismanagement, lack of community consultation, broken agreements, and victimization of residents.

There is an unprecedented and overwhelming intrusion of violence and crime together with antisocial behaviour in this community, perhaps unmatched anywhere else in Australia on a per capita basis. Application to support this claim has been made under the freedom of information act to the relevant authority and access to these figures and statistics has been denied.

The victimization of community members is at all levels, especially the vulnerable. The elderly, infirm, women and children and young families have all suffered. In addition established businesses have been robbed, threatened and trading affected by this situation. Community standards of public safety and quality of life are at a new low point in the area despite an enhanced police presence.

The undersigned now want specific performance from our parliamentary representatives, the Premier and Dr Refshauge with the establishment of a permanent police station on the Wilson Brothers factory site. This site, now South Sydney Council Property was vacated by an established business due to relentless crime and victimization.

Citizens will welcome such a police presence, and the re-establishment of law and order in the area, as will established businesses and other bodies such as The Australia Council, Sydney University and the Australian Technology Park.

As citizens, ratepayers and voters, we the undersigned demand a return to proper standards of law and order in our community through the expeditious establishment of a Police Station at the Wilson Brothers site.

My personal experience of crime in this area is:

Just another day in Redfern

A SPATE of bag snatches in the inner-city yesterday — in which six women were robbed or assaulted in separate incidents — was "just a normal day" for the Redfern area, police said.

Five reports of bag snatches were made to police between 10am and 1pm, with another later in the afternoon.

Some victims were assaulted and descriptions by witnesses indicated at least two of the crimes were committed by the same person.

But the seemingly high number of bag snatches was described by one Redfern police sergeant as "just a normal day".

"It wouldn't be a surprise to see someone walk in here now and report a bag snatch," he said after the sixth incident had occurred yesterday.

— THE DAILY TELEGRAPH, Monday, April 20, 1998—7

Figure 3.2 'Just Another Day in Redfern' Petition

profile television reporter and anchor, Kerry O'Brien, at the ABC TV 7:30 Report.

Several weeks after the door-knock, and the 'petitions' being sent to the 7:30 Report, the 'Manager Editorial Development & Support' for the program responded with a 'Dear Resident' letter that was sent to all residents who participated (or, as was the case with my household, had NOT participated!). This letter was a polite refusal to pursue the story. Anti-Block activisms have been, at times, manipulative and, in the case just described, legally suspect.

The laneway at the heart of the aforementioned furore remains a contested and highly racialized zone between what appears to be the agendas of 'black' drug users versus 'white' homeowners. When the 'needle-exchange' bus visited the laneway, there were two consequences. First, essential health services were provided. The second consequence was that it helped to formalize the status. Drug use had, by association and proximity to The Block, become an Aboriginal issue regardless of the supply and the demand: this mobile needle-exchange facility bene-fited both Aboriginal *and* non-Aboriginal intravenous drug users, and had it not existed, that 'white child' may have been exposed to disease through use of a second-hand needle and syringe. Non-Aboriginal heroin use, so commonly 'an issue' in other inner-city areas, remained hidden because of the racialization of drug use in the Darlington/Redfern area. Many believe that the government can somehow remove *all* drug-related activities from the area or, at the very least, they should remain confined to The Block.

This next section details another example of the way that The Block, drugs, and the seepage of Aboriginal drug use into neighbouring areas, have been countered by resident activism. Since the aforementioned media report, and the removal of the mobile needle-exchange facility from the laneway in 1999, the NSW Minister for Health and Aboriginal Affairs announced the establishment of a number of new NSW Health Department initiatives in the Darlington/Redfern area. One proposal was to house a facility that would target drug dependency in a building in Little Eveleigh Street, near the corner of Lawson and Eveleigh Streets (RRAG 'letter to sign and send' 17 February 1999, *Chippo Politics*, 2, 2, March 1999). A group of householders responded to the facility pro-posal by rallying to the battle cry of 'not another shooting gallery'. The (now defunct) Redfern Resident Action Group (RRAG)[25] formed imme-diately 'to voice the concerns of local residents about drug dependency services in the Redfern area and to offer solutions through proper community consultation' (Redfern Residents Action Groups Position

Paper, 1 March 1999).[26] The first RRAG missive was letterbox-dropped around Darlington, which stated:

> A recent internal review by the [NSW] Health Department on the operation of the injecting equipment vehicle in [the laneway] was released to the media on 7 February 1999...recommend[ing] the placement of a large scale permanent drug dependency service near The Block [leasing] 122 Little Eveleigh Street, to set up a needle exchange program in that building. (from RRAG 'letter to sign and send' 17 February 1999, also distributed at first public meeting)

The aim of the RRAG campaign was to halt the development of this 'large scale permanent drug dependency service near The Block'. The rationale was that such a facility should not exist in a 'residential area', and specifically not in a non-Block part of Darlington. The campaign started with urgent public meetings (starting February 1999), and the media followed the story 'Residents fear Redfern clinic will be needle exchange' was published in Sydney's daily broadsheet (*SMH* 17 February, 1999). The 'sign and send' letters (see quote above) were addressed to various politicians, including the NSW Minister for Health who was also the NSW Minister for Aboriginal Affairs *and* the Deputy Premier of NSW, at that time. The politicians targeted by the 'sign and send' letters were invited to attend and speak at public meetings. Most did attend, and most addressed the meetings (from local, state and federal governments).[27] The local area police commander also addressed a meeting and professed his support for those residents who opposed the facility. Newsletters kept local residents informed of the progress of the campaign, and the success of lobbying against the facility.

Although couched in different terms, the RRAG response to the proposed health facility was not dissimilar to an earlier proposal for the imposition of a noxious facility. This familiar 'Not In My Back Yard' response also paralleled the response to a proposal decades earlier: to house an Aboriginal community in the area. Michael Dear's (1992) 'good neighbour hierarchy' (rather than the implied 'bad' neighbour hierarchy) included groups with physical disabilities or life stage issues such as 'old age' or 'terminal illness' at the top of the list of tolerances. Lower down the scale were facilities for 'mental disabilities', and at the bottom of the hierarchy of acceptance could be found services for those with 'social diseases' such as criminality, alcoholism and drug use.[28] Regardless of the general applicability of this hierarchy of acceptance, the aforementioned responses in the Darlington/Redfern area exposes how located processes, such as extreme racialization of a particular group, can significantly shift any standards in a hierarchy of acceptance (cf Pendall 1999). The proposal to house an Aboriginal community in the 1970s was met with responses that were usually reserved for facilities at the bottom of Dear's hierarchy. Moreover, with gentrification in inner Sydney, the demographic shift towards a more affluent populace has meant that tolerances to existing welfare facilities have also decreased.[29] The campaign against the proposed Little Eveleigh Street health facility was highly organized and well resourced. It drew support from a wide and influential field.

The influences of highly mobilized resident action on government process are worthy of mention, particularly with such a highly charged example as this. In the early stages of the campaign for the Little Eveleigh Street health facility, the NSW Minister for Health and Aboriginal Affairs largely ignored the pleas by RRAG to reconsider the Little Eveleigh Street proposal. At that time, one Independent MP, Clover Moore, did listen and responded to the residents who were being represented by RRAG. Electoral boundaries had moved and the Darlington/Redfern and Chippendale neighbourhood had suddenly come into her electorate. She had promised 'to make inner Sydney a better and safer place to live' ('letter to residents' leaflet from Clover Moore's electorate office, 25 February 1999), and sent a letter to the Minister. It was titled 'Re: Proposed consultation on drug and alcohol services at Redfern' and was also letter-boxed to residents of Darlington (dated 25 February 1999). In this letter, Moore claimed to represent 'the residents' who were 'highly sceptical that an authentic, meaningful and effective consultation process will take place'. The Mayor for South Sydney City Council, who was competing for Moore's seat in the state parliament at that time, quickly added his support for RRAG and their campaign. The desires *not* to have this facility 'without the support of the Redfern community' (letter to 'Residents' from Vic Smith, Mayor of South Sydney, 23 February 1999) became a political competition. RRAG successfully timed their campaign by pitching it at rival political parties just before an election. The battle, to establish which political party would best represent these residents, overshadowed the real issue of the drug problem. The strength of resident activism was realized as the proposal for a facility that might help to address a problem that was largely identified with The Block in the Redfern/Darlington area, had faltered. The residents had won that battle, for the moment.

The RRAG campaign was also assisted by some unsolicited support. The term 'shooting gallery' had been a part of the local lexicon, particularly since the laneway 'incident' discussed earlier. This emotive term continues to conjure images of orgies of heroin use and guns that have been depicted in films about US 'ghettos' and drug taking. An unnamed group quietly exploited the situation with a leaflet titled 'Are You Nervous As You Approach Or Leave Redfern Station', which was dropped into Darlington letter boxes, while poster-versions were pasted on telegraph poles around the neighbourhood (Figure 3.3). This poster/leaflet tied the Little Eveleigh Street facility directly to The Block, via the aforementioned laneway drug issue. RRAG quickly dissociated itself from this publicity for fear of being associated with this pointed reference to The Block, and the possibility of tarnishing their campaign with

---

ARE YOU NERVOUS AS YOU APPROACH OR LEAVE REDFERN STATION ?

The Government statistics for our area say you should be with its incredible crime rate: see http://www.lawlink.nsw.gov.au

Well, its only going to get worse with the new Vic Smith-sponsored shooting gallery alongside Redfern Station in the red brick Association for Good Government building in Little Everleigh St. The lease has already been signed in secret without any community consultation whatsoever.

This shooting gallery will directly affect you - your safety, your property, your children, your lifestyle. Last year 500,000 needles were given out in nearby Caroline Lane.

Letters count, so write and voice your opinions to Andrew Refshauge, Deputy Premier, Minister for Health and Aboriginal Affairs Level 11 - 73 Miller St - North Sydney 2060 Telephone 9391 9888 and Fax 9957 2145.

Support the Redfern Residents Action Group

Figure 3.3 Are You Nervous? Leaflet

what might be construed as racism. RRAG maintained the position that the medical service simply *should not be in a residential area*. It was easy for the RRAG campaign to not speak in racialized terms – the unnamed support group had boldly stated the message that the RRAG had politely, or strategically, skirted.

Again, the issue of non-Aboriginal drug use had been avoided as the focus remained firmly on the mostly unspoken but assumed association with The Block. When non-Aboriginal people engage in heroin use in the area, such as the 'white child' depicted in the laneway story, drug use is still associated with The Block. In 1999, the NSW Health Department released a report on results of community consultation about 'Drug services in Redfern' (carried out by a private consultancy company). The report concluded that there was 'widespread agreement that drugs and alcohol issues...represent a major health and social problem in Redfern...On the other hand [there are concerns that] any significant development of services in the Redfern area...[are] likely to enforce Redfern's role as a centre of illegal drug dealing and using in Sydney' (Health Department of NSW 1999, 3.1). Throughout the report, the linkages between the 'drug problem' and The Block were simply assumed.

Years late, in June 2005 a public meeting was organized (and paid for) by a new resident action group, REDAlert, in the Redfern Town Hall. REDAlert formed to oppose yet another state government proposal for a drug-related health facility in Redfern. The group preceded the meeting with a graffiti campaign (Figure 3.4), and over 150 people, including journalists and film crews from ABC TV, 2GB radio, and others, packed Redfern Town Hall for what became a highly emotive event.

The baby-toting MC of the meeting, and convenor of REDAlert, opened the meeting with the immortal words 'this is about family!' The hall was filled with balloons, and babies and toddlers were in abundance. The walls of the hall were plastered with large imposing posters that read 'No Needles Next To Children', and sensationalist newspaper clippings that had been enlarged for easy reading carried headlines such as 'It's not Mr Whippy [ice-cream van], it's the Needle Van' (*SMH* Editorial, June 2005, 26), and 'The last thing Redfern Needs' (*SMH*, Miranda Devine, 17 June? 2005). Overwhelmingly, the meeting reflected a familiar 'anti-facility' position.[30] The events surrounding the failed Little Eveleigh Street facility were echoed and the fate of this newly proposed facility hangs in a similar balance as the latest round of political football is played. Meanwhile, the inner-city drug problem remains unaddressed.

Any proposal by governments or any organization for that matter, for a drug-health facility in the Redfern area will be met with hostility. Any

Figure 3.4   REDAlert Graffiti Campaign

threat of (imagined) seepage of drug use (and disease) along the notorious laneway, or across Lawson Street and into non-Aboriginal residential space, would meet with resistance. Issues of control, at racialized housing borders, are complexly articulated and enforced, particularly where behaviour is deemed inappropriate or aberrant (Sennett 1976, Gregory 1993, Smith 1996). Drug use and related crime is a serious problem in the Darlington/Redfern area but any attempts to alleviate this problem, such as by the provision of a specialist state-run health facility, continue to collide with the agendas of increasingly intolerant resident activists. The legacy of the original objection to the formation of The Block (by the SSRPM in the late 1960s) remains. The prejudices have become entrenched, as fears about The Block, which are now associated with drug use and supply, escalate. This escalation will not be alleviated by the provision of a health facility that might have an impact on this issue. Instead, such a facility would appear to be yet another provision for Aboriginal people to enjoy, at the expense of (non-Aboriginal) taxpayers. The provision of such a privilege would be highly resented, and will be stymied at every turn. In the next example, I detail how the future use of a site that had been earmarked for housing

Aboriginal services was successfully contested to ensure that (the myth of) Aboriginal privilege would not occur in Darlington. Instead, a process of whiteness, through the fortification and consolidation of gentrifying space, was successful against yet another imaginary threat of Aboriginal seepage from The Block through a surreptitious grab of yet more Aboriginal territory, by the unacceptable Aboriginal 'other'.

Seizing space: the unruly fortunes of the Wilson Bros. site

> At the time of preparation of the plan the buildings are *generally vacant* and derelict. The Redfern Aboriginal Corporation[31] runs employment training programs on part of the site and is the only current user of the land. (Section 2.3 Current Uses of the Site, South Sydney City Council 1999, 1, *my emphasis*.)

In another example of the interplay between local resident activism and government, the fate of the Wilson Brothers factory site (herein Wilson Bros. site) had serious implications for the Aboriginal community on The Block. The Wilson Bros. site adjoins The Block, and in 1990, the South Sydney City Council (SSCC) endorsed a plan to develop it as an Aboriginal recreational centre. In 1993, the SSCC purchased the site (bounded by Caroline, Hugo, Vine and Eveleigh Streets) from Wilson Brothers Pty Ltd. as part of a wider redevelopment plan for The Block (SSCC 1999, iii). The site, which consisted of four large factories and several small terrace houses, was consequently re-zoned from industrial to recreational usage. Several improvement plans for The Block were in place at the time, and the Wilson Bros. site would consolidate the provision of Aboriginal services within the immediate vicinity of The Block.

However, by 1997, the site remained undeveloped and trouble erupted when SSCC announced its plan to demolish the site's buildings. According to Section 2.6 of South Sydney City Council's 'Draft Plan of Management of The Wilson Bros. site and Yellowmundee Reserve, Caroline Street, Redfern' a decision had been made to turn the site into open space, temporarily, in response to 'concerns regarding the structural safety of the buildings' (SSCC 1999, iii). The SSCC had a new proposal: to turn the site into green space.

With the release of the (new) SSCC plan, concerns were suddenly raised about the possible heritage value of some of the factory buildings on the Wilson Bros. site. Another concern was that usable buildings were to be demolished. The Redfern Aboriginal Coalition (RAC) leased one of buildings at 'peppercorn' (low) rent. The RAC provided valuable employment schemes for the local Aboriginal community. When the safety of existing buildings was questioned,[32] the RAC engaged engineers to check the building that they were using and it was deemed to be safe (RAC, Pers. Comm., December 1998). Regardless of concerns about safety, there was enough opposition to the proposed demolition for SSCC to take the course of consulting with 'the community' to assist in the decision about the future of the Wilson Bros. site (SSCC 1999, iii). SSCC engaged a private community consultancy company to 'identify the individual people and groups with an interest in the development of the site to allow them to participate in the preparation of the plan' (SSCC 1999, iii). After 35 weeks of 'community consultation', carried out between November 1998 and July 1999, a draft 'Masterplan' and 'Draft Management Plan' was presented to the SSCC.

The final plan recommended that all but one of the large factories, deemed to have heritage value, be demolished. The remaining building would be restored, and used for 'General Community Use', and one small corner terrace house would remain.[33] The rest of the site would become parkland (SSCC 1999, 42). The 'recommended management scenario' in the Draft Management Plan (SSCC 1999) was that the SSCC should act as a sole manager for the site in order to establish the centre and, its community programs. The SSCC would grant leases and the centre manager would carry out day-to-day administration of the leases (SSCC 1999, 48, Section 8.4.4).

One glaring omission to the plan was the future of housing Aboriginal services, such as the RAC, or a recreation facility. On this matter, the SSCC had been 'not very helpful' according to the spokesperson of the RAC (RAC, Pers. Comm., December 1998). The RAC has since had to relocate to a more expensive, and remote site, a kilometre or so away from The Block[34].

The following is an account of how the process of public participation that led SSCC to abandon its original plan to house Aboriginal services on the Wilson Bros. site played out some of the now-familiar strategies of whiteness, of siege-based reactions to the fear of the Aboriginal 'other', that have become commonplace in this part of Sydney.

Community consultation?

The first public meetings in the community consultation process were held in November 1998. These meeting were emotionally charged.[35] Some residents used these meetings as venues to air grievances about perceived drug use and crime in the local area. At one meeting, in Ivy Street, The Block community was repeatedly blamed for crime in the area. The two Aboriginal Elders were clearly distressed by the overt hostility directed at them. A consequence of this kind of experience was that the Aboriginal presence at the consultation sessions diminished very quickly.

At the final stage of the community consultation process, in May 1999, a public meeting was organized. Three options for the Wilson Bros. site were finally presented and 'the community' was invited to contribute by *choosing* a preferred option. Option 1, 'Building and Courtyard' proposed to retain the main buildings around a central courtyard. Option 2, 'Building and Park' proposed to retain a (heritage) building at the Northern end of the site, demolish all other buildings and convert most of the site into a park. Option 3, 'Park and Place' proposed to demolish all buildings and convert the site into parkland.

Box 3.1   Wilson Bros. Factory Site ~ community consultation*

Aim: to decide what to do with the site.

Process: Lengthy public participation.

1) Prospective steering committee members nominated
2) Council selected committee
3) Public meeting process (private consultancy group)

Public Meetings:

1) Stage 1 Public Meetings: 4 separate meetings (all 'the same')
2) 'Speak Out' 28 November 1998 (Wilson Bros. site). (Comments from the 4 initial meetings displayed and commented on)
3) 'Checklist Workshop', 6 February 1999 (Darlington Activity Centre), where 'a set of principles to guide the planning of the site ... workshopped and refined ... a draft set ... presented based on results of the previous consultations'
4) 'Nitty Gritty Workshop', 8 May 1999: 'a public participatory design workshop where plans showing alternative options ... worked within a 'hands-on' way by the community. The 'check-list' ... used to assess plans for the site that satisfy the needs and aspirations already expressed by members of the local community'.

* Summary of SSCC Community Participation Newsletter, January 1999.

Options and stakeholders

From the early concerns about the structural safety of the Wilson Bros. site buildings and the announcement that the SSCC would demolish the buildings, to selection of the 'Steering Committee', through to the final three site proposals, resident action groups mobilized. The 'Steering Committee' was selected by the SSCC to 'provide... direction for the study and operate... as the principle conduit for contact with the community in the interim periods between formal community events' (SSCC 1999, 11). The Steering Committee consisted of representatives of four main stakeholders, as well as three council representatives. The stakeholders included the Laneway Resident Action Group (LRAG), the Redfern Aboriginal Coalition (RAC, as distinct from the Aboriginal Housing Company – the landlord organization for The Block – which was not involved in overt lobbying), Redfern Residents for Reconciliation (RRR), a small group of Aboriginal and non-Aboriginal residents, and the residents for Wilson Bros. (RWB)[36] (consisting largely of KAK members). A representative of each group was selected for the steering committee and the process of public participation resulted in three options (Wilson Brothers site Redfern, Newsletter # 2, May 1999, South Sydney Council).

- **Option 1**, 'Building and Courtyard' was proposed in line with what the consultants believed to be the desires of the RAC, and supported by the RRR. The RAC wanted to retain buildings for Aboriginal services, including their own. This option was *not* endorsed by the RAC because it did not represent their case for the provision of accommodation for (existing, and new) Aboriginal services. According to the South Sydney Council's Newsletter (Wilson Brothers site Redfern, Newsletter # 2, May 1999, South Sydney Council), 'the following uses could be accommodated... children and youth services, community services, activity rooms, multi-purpose hall, workshops, support facilities'. The words 'Aboriginal', 'Indigenous', or indeed 'employment' were not included with Option 1.
- **Option 2** proposed to keep the building that some residents felt had heritage attributes, for a sports/community hall, and convert the rest of the site into parkland and recreational areas. This option did not appear to 'side' with those who wanted open parkland (and no housing of Aboriginal services), associated with Option 3 and its supporters, nor did it 'side' with those who preferred Option 1 (imagined to be the preference of the RAC and the Aboriginal

community), to keep buildings (for lease) and have a central courtyard.

- **Option 3** was the preferred option of RWB. It required the demolition of all buildings and the construction of a park. The resident
'feedback' about this option was that it was the 'safest' option
(Wilson Bros. site Community Participation Program Stage 2, 28
November 1998). Narratives of dirt, danger, drugs and crime (associated with The Block), peppered the feedback documentation
('Speak Out' feedback, 28 November 1998, SSCC). 'Open space'
was preferred because it would allow better surveillance of the
Aboriginal community. As one resident stated: 'We just want a
park. It [the site] goes half way down to Vine St, [so] it's easy to
police as well. Most people here, if someone says boo you call the
police' (Interview, 13 Sept 1998).

Within the campaign for a park, a sub-campaign for a police station was
mounted but was swiftly quashed by the Local Area Commander of
Police who made it clear that there would be no police station housed on
the site. The proposal for a park *that would provide open space for
surveillance purposes* became the agenda of the campaign. The first
theme for this campaign was an evocative plea for 'green space' (a
combination of park and open space) in the congested city. At the first
public meetings, a representative of RWB presented their case in this way:

> The best solution for the [Wilson Brothers] site is a Native Australian
> Flora park ... because ... it is an open valley of natural Australian trees and
> bushes. A return to the pre-colonial and possibly even the pre-aboriginal
> [sic][37] landscape of Redfern. Can you remember the feel of the Australian
> bush? The smell, the light, and the sounds which have long disappeared
> from Redfern? ... It's fresh ... It will attract native birds. Can you imagine
> hearing the raucous laughing of kookaburras in the inner-city? It *will*
> happen. And it will bring Cockatoos, Rosellas, Galahs, possums, and
> magpies ... It prevents the site from becoming another failed social experi-
> ment in Redfern. ('You can help decide the future of Redfern', Leaflet,
> 17 November 1998)

This seductive narrative of bushland in the city was used for two purposes (at least). It conjured an imaginary of a breath of fresh air in an
otherwise congested inner-city area. Second, this pre-Aboriginal (as well
as pre-colonial) bush imaginary invoked a fantasy natural landscape,
which in this case was free of Aboriginal people, pre-colonial or otherwise.

This version of a pre-Aboriginal park was invoked only the once, and thereafter, the focus shifted to the provision of 'open space' to provide 'visual links' (SSCC 1999, 40). This surveillance, the capacity to watch the Aboriginal community (and thereby gain more control of the 'Aboriginal (drug) problem') became the new catchcry of the RWB.

The draft 'Master' plan of management presented a more neutral interpretation of the visibility that open space would deliver once buildings were demolished:

> A community precinct linked visually and physically to the local area...
> accentuates existing visual links across the locality so as to facilitate social
> interaction both at the site and through the neighbourhood. (SSCC
> 1999, 40)

The term 'visual links' represented a softening of surveillance agendas, and was repeated throughout the final Draft of Management Plan document (SSCC 1999). One resident, who worried about the proximity of The Block to the Wilson Bros. site, offered a different perspective on the notion of 'visual links':

> I don't want a park because it will be noisy. There will be no sleep at night.
> I would like townhouses because it puts more value on our property. No
> park!...People will be drinking and drugging all night. ('SpeakOut'
> feedback, 28 November 1998, SSCC)

This resident was clear about a fear of the future usage of the Wilson Bros. site, that it be redeveloped as housing, or pay the price of open access to The Block (and vice versa). The usual problems associated with a proposal for development, the increased traffic, and parking difficulties and, in this case, the issue of the loss of publicly owned council land, were not *the* problem, for this resident.

Finally, SSCC rejected Options 1 and 3 because:

> [I]t was considered that neither adequately addressed the expressed needs
> of the community for development of the site...the 'building and park'
> option [2]...is considered to best address the express needs of the local
> community and best respond to the characteristics of the site and its
> surroundings. (SSCC 1999, xiii)

Option 2 appeared to be the fairest and most neutral choice for council to make within such a heated and contentious debate.

The politics of silencing: 'Option 4'

At the final community consultation workshop, which was the forum at which the final three options were presented, an alternative option, the 'Redfern Reconciliation Community and Cultural Centre' option, dubbed 'Option 4', was presented by a group of participants who felt that Aboriginal voices had been silenced. This 'shared vision' was the result of meetings between the Redfern Aboriginal Corporation (RAC), The Settlement (a local Aboriginal welfare agency), Redfern Residents for Reconciliation (RRR) and Redfern and Environs Neighbourhood Education Workshop (RENEW). 'Option 4' included architectural plans for improving the existing site and also planned to retain much-needed accommodation for Aboriginal services, including the RAC. It had been offered to the community consultants before the workshop but was rejected for reasons that remain unclear. So then, 'Option 4' became a protest. Supporters silently held hand-painted signs that simply stated 'Option 4'. The team of consultants and landscape architects had not prepared this option.

The point of the protest was that the 'community consultation' process had been flawed; it had marginalized the Aboriginal people associated with The Block. The consultants verbally reiterated a position that no particular stakeholder was going to be seen to have a majority say in the future use of the Wilson Bros. site. It seemed that the SSCC's original intention, and reason for purchasing the Wilson Bros. site to house an Aboriginal recreation facility, had been completely sidelined to this 'position'. The consultants had decided to disregard those in support of 'Option 4' in the same way that those who were overtly hostile to Aboriginal people were also disregarded. Both 'groups' were treated as two opposing factions, as protesting fringe groups.

'Community becomes uncivilized'[38]

During the final community consultation meeting, a leaflet was distributed by the RWB. This leaflet had also been letter-boxed around the Darlington area, prior to the meeting. In response to this leaflet, an Elder from The Block gave an impromptu (and moving) speech on racism, which brought the leaflet and its racializing agenda to the attention of those present at the meeting (see Figure 3.5).

Far from representing all sectors of 'the community', particularly the 'powerful Aboriginal lobby' or the RAC, as claimed by the RWB (Figure 3.5), the consultation process had eliminated the possibility that the site

_____ GROUP

NITTY GRITTY WORKSHOP
MEETING AT THE SETTLEMENT
8 EDWARD STREET CHIPPENDALE
SATURDAY 8th MAY 11:45AM – 5:00PM

This is the last public meeting regarding the Wilson Brothers site before a decision is made by council regarding its use. We really need as many people as possible to attend and put their views forward. Please make the effort and come to the meeting, if only for a few hours. It will make a difference to this area.

PURPOSE: This meeting allows you to help design the site – to move trees or facilities around on a scale model. It also allows you to have a final say in what you want for the site.

We appreciate all the support we got to get one of our representatives on the council steering committee. We are one of 4 community representatives on the said committee and a lone voice for at least 400 people supporting us.

The fate of the Wilson Brothers site will be decided based on community opinion. If we cannot show exactly what the majority wants, how can we get our point across?

Don't think that your contribution won't make a difference. IT WILL!!!

There is also an email address where you can put your view forward. It is:

Remember, this is your chance to make a difference!

The view of _____ Group

Proposals for the site generally fall into two polarised views.

There is one group of people who are proposing that the site should become workshops and offices principally to house the Redfern Aboriginal Corporation (RAC) and other "community" events. This group is mainly represented by the RAC and the RRR (Redfern Residents for Reconciliation). They have some support from residents and a number of other aboriginal groups in what amounts to a powerful "Aboriginal lobby".

We (the _____ Group) have been campaigning for a long time to turn the site entirely into a park. We have been doing this only because we believe that we represent the views of many residents who would otherwise be ignored. We have no special interest groups to help present our case.

When developers apply for construction of units they are required to pay a levy to council which must be used by council to provide open space for recreation. South Sydney council has used this fund to purchase the Wilson Brothers site. Hence it is zoned 6A recreational - community open space. Change of zoning would be almost impossible and would probably be strongly opposed in court by the developers who contributed to the fund, as well as residents. The zoning does not allow for commercial activity.

We have been campaigning to defend this zoning, both in principle AND in actuality, and in doing so we are defending the quality of life of all residents in the area.

Certain people seem to believe that the local aboriginal community have a natural right to this land, that the site should be a place for aboriginal culture, and that the site should represent a symbol of reconciliation. We do not agree with any of these views. The site has nothing to do with indigenous land rights or reconciliation. It is community park land.

The RAC has released their proposals for the site, which consists of a number of buildings around a centre, courtyard all surrounded by a fence. The plan is designed with only their own activities in mind. Despite the fact that they have expressed intentions to include the whole community, their plan shows otherwise. In fact many long-term Redfern residents are shocked at their plan. We believe it is a greedy land grab for the RAC to expect Council to give them something they have absolutely no right to. They must find other premises for their activities, and not assume that this is their site. The site belongs to all residents of the area and you are defensive that principle.

A middle ground plan proposes that approximately one-third of the site remains a building and the other two thirds become a park. There is only one reason why this proposal has been put forward; *it is a compromise*. Other than that it doesn't help either group. It proposes that the remaining building be converted into a community hall. There are already enough community buildings, meeting halls, etc in the area. We really do not need any more. What we DO need is more recreational space. Every function that has been suggested for use in this community hall can be accommodated easily in other premises in the surrounding area.

We ask you to reject both of these proposals because:

• They go against the zoning, which specifically states that the site will be used for recreation -community open space.
• They achieve nothing for the wider community
• They rob us of the space we need to develop a significant contribution to the revival of 'once city green space'
• In all reality it will just become another set of vandalised buildings.

It is a step backwards. We are sick of what's been happening in Redfern over the past two decades, where every issue is dominated by so-called aboriginal welfare and we have nothing to show for it except a poor quality of life characterised by violence, fear and drug-addicts. Clearly the methods of the last twenty years have failed and we do not want an expansion of the same old things

We support the principals of the RAC and we support any measures which will improve the quality of life for the genuine aboriginal people of Redfern. We believe that making the entire Wilson Brothers Site into a park will, in the long run, most effectively achieve these aims.

HOW CAN WE BE CONTACTED? If you want more information about how we formed and more explanation about what we are trying to achieve, or you want to make some contribution or criticism, please contact us.

1. Leave a note at the _____
2. Contact us by email. Our address is:

HOW TO HAVE YOUR SAY:

1. Attend the Nitty Gritty Workshop. We ask you to support the park only option
2. Sign the form below and either drop it into the Nitty Gritty workshop or mail it to
   South Sydney Council
   ATTN: Laurie Johnson
   Public Works and Services
   140 Joynton Avenue
   Zetland NSW 2017
3. Email your support for the park only option to

I am(we are) local residents (or interested party) and support only the option for the Wilson Brothers Site to become entirely a park.

Additional Comments: _____

Name(s): _____

Address: _____

Signature(s): _____

Figure 3.5   Nitty Gritty Workshop Leaflet

would become a recreational centre for The Block. This original purpose, the reason that the SSCC had purchased the site in the first place,
was not mentioned at *any* of the community consultation meetings.
I found this information about the purpose of the purchase in Sections 2.6
and 2.7, both titled 'Recent Planning History', of the Draft Plan of
Management (SSCC 1999, iii, 7). This Draft Plan of Management
document was made available *after* the community consultation process.
Why this information was omitted from the 'community consultation'
process remains unclear. What is clear, however, is that even if this vital
piece of information (vital enough to be included in the Draft Management Plan) had not been known to the consultants at the start of the
process, it was certainly was brought to their attention *during* the
community consultations.[39] The result of that process and the final
plan for the Wilson Bros. site may have been very different had 'the
community' known of this original purpose.

The draft Plan of Management recommended that 'the park component' be implemented at the outset of the redevelopment project. Although
this aspect of the project occurred very quickly, with the wholesale
demolition of buildings, which cleared the way for surveillance purposes,
this was not enough for some. According to two residents:

> Yeah we tried to close Caroline Lane, and that didn't work, people
> objected. Now...The Wilson Bros. site...They [the Aboriginal commu-
> nity] want to keep it. Council owns it, so we're just against any extension
> of The Block. We want it to be a park. We want part of it to be a police
> station as well, and we got over 400, no 390 signatures...And some [of
> the 390] people objected to the police station. That was the biggest piss off
> because they didn't have any other ideas, they didn't say why. (Interview,
> 11 September 1998)

> ...the 500 letters we got together prior to the steering committee, were
> just completely ignored by the 7:30 report, by the Prime Minister's office,
> by Bob Carr's office [NSW premier], by the police commissioner, and so
> on and so on...And then they turn around and say, we're going to put a
> shooting gallery [needle-exchange] in [at The Wilson Bros. site], yeah, its
> pretty frustrating. (Interview, 19 September 1998)

Non-Aboriginal residents who live in close proximity to The Block tend
to respond to its existence in one of two ways. On the one hand, there
are the anti-Block agitators who maintain the rage against it. This group
mobilize support by focusing on the unruly aspects of living near such

a disadvantaged group. They actively recruited newcomers, and are adept at reinventing their collective identity, as required. The RWB (known also by various other names) was an offshoot of KAK, which formed in 1997–8 in response to the issue of crime in the local area. Those 390–500 'victims of [racialized] crime' stories mentioned previously, appeared repeatedly, in various campaigns. On the other hand, the local branch of the Redfern Residents for Reconciliation (RRR) formed to support the RAC. And there are those that continually show their support for The Block. Some lobby governments, some attending rallies, and some just live their lives and accept that they are in the company of the community on The Block.

The activisms that have emerged in response to the fear of the 'other' have relied on maintaining this fear. Campaigns that have focused on Aboriginal drug use and criminality have relied on this strategy. Agitators from KAK (and its other incarnations), and their attempts to (re)-produce fear, may have failed in their quest to have a police station on the Wilson Bros. site but they did succeed in their mission to open spaces of surveillance or 'visual links'. An outcome of the Wilson Bros. site community consultation process, and consequent decisions made regarding the site were ultimately, however, far worse for the local Aboriginal community than the exacerbation of non-Aboriginal fears about their presence. The geography of fear, which may unsettle non-Aboriginal residents, was one thing. The erosion of *material resources* that resulted from the production of fear has remained largely unacknowledged.

In this example of the locally contested Wilson Bros. site, any merging of community interests was subsumed by what became a racial(ized) contest. Far from bridging the racial binary, the final 'compromise', Option 2, simply reinforced community division. In its attempts to be inclusive, the SSCC made provision for a process that was, in the end, exclusionary. The consequent process enforced an existing power relation in the Darlington/Redfern area. Any possibility of 'Aboriginal privilege' (Mickler 1998), such as the provision of an Aboriginal recreational facility and the ongoing accommodation for the RAC, was averted.

Protecting neo-colonial heritage

Apart from KAK and its various *alter egos*, and the groups that formed around the Wilson Bros. site consultation process, the other kind of resident activism gaining strength in the Darlington/Redfern area is the protection of built heritage. I noted in the previous chapter that the question of which pasts are remembered and which pasts are forgotten or denied rarely arises during considerations of what constitutes heritage and its protection. In this final example of resident activism, a more nuanced process of what I will call 'whiteness' has emerged in the name of heritage.

> There is...a nexus between the identities of places and the struggles between the social groups which seek to create, or to modify, these identities and it is this nexus which so often leads to the juxtaposition of such terms as 'Dissonant'...or 'Contested' with the word 'Heritage'. (Shaw and Jones 1997, 2)

Just a few streets away from The Block, another resident action group (that I have dubbed Resident Action Group, or RAG)[40] formed in early 1997. Its stated purpose was to protect Darlington's built heritage. However, a close investigation of the RAG campaign has revealed quite a different style in mobilizing whiteness. In this example, the link between efforts to protect heritage and an exclusionary neo-colonial identity politics can be found. This is a very different politics from the reactionary resident activism discussed earlier. This kind of politics uses its investment in the cultural capital (Zukin 1995) of heritage buildings and architecture, and taste-based consumption (Bourdieu 1984), to determine the appropriateness of residential development. In what follows, I will trace how RAG was able to flex a political muscle as a group of unabashed new gentrifiers who actively sought to reflect their exclusionary value systems through architectural expression.

New developments in the Darlington/Redfern area are increasingly required to respond to the existing heritage character. Taste-based determinations of heritage are closely negotiated between local government (formerly SSCC, now Sydney City Council), property developers and more recently groups such as RAG. During the process for gaining development approval for a warehouse conversion in Darlington, RAG successfully prescribed for the exclusion of low cost accommodation for students (or others in need of affordable urban accommodation). RAG members, led by resident architects and designers, successfully negotiated themselves into the council processes that determine who – that is, the kinds of people – were acceptable residents for a corner of the neighbourhood.

Architects of taste

> Architecture and structure is one of the major physical attributes of our
> past that we bump into daily... there needs to be a lot more concern and
> consideration for things of the past... it's really important to hold and
> admire and to talk about, otherwise we're denying our past... [so] we
> convened a group called RAG. (Interview, 4 February 1998)

In December 1996, a 'tasteless' residential development (Interview,
4 February 1998 and RAG campaign newsletters) had been proposed
for a site in Darlington that is only a few streets away from The Block.
The developer planned to 'demolish existing buildings and erect a 2–3
storey building with 45 [home] units and 32 basement car spaces' (SSCC
letter to residents regarding development). A small warehouse and a
larger factory were earmarked for demolition. This proposal generated
a protest that received widespread support from residents. Public meetings were well attended and RAG mobilized a petition in objection to the
Development Application (DA) (submitted to SSCC). The rationale for
the objection was that the new development threatened the 'heritage
character' of the local area, and the existing diversity of the built
environment (Interview, 4 February 1998).

Darrel Crilley (1993) has raised suspicion about the use of the term
'diversity' in the promotion of architecture. Because architecture is often
used to promote or 'advertise' meanings, Crilley (1993) has identified
that such meanings can include the promotion of diversity to enhance
the *appeal of* developments, rather than to *appeal to* a diverse market.
In the UK, John Gabriel has written of the effects of 'whitewash', as a
way to conceal homogeneity within the *appearance* of diversity. 'Diversity' is often simply another consumable attribute for affluent tastes
(cf Hage 1998, Bourdieu 1984) and rather than appealing to a range
of types of people, only those with the necessary attributes (such as cash,
class and/or ethnicity) can have membership in such a niche market.

It was the threat to the architectural diversity of industrial buildings
that mobilized RAG's campaign: 'we managed to save the façade of
the... warehouse... through 6 months work... we got what we
wanted, in a way' (Interview, 4 February 1998). The issue of preservation, for RAG, reiterated the themes of 'façadism' and exclusive/exclu-
sionary designations of heritage discussed earlier in this chapter. Far
from a desire to protect the heritage of human diversity (such as the
memory of those who toiled behind the façade – the poor 'working

classes' who consisted of a range of ethnicities), the issue of *who* would occupy the new development was of greater concern at the RAG meetings. There was the possibility that the new residential development would attract (overseas) students because of its proximity to two universities and various residential colleges (Public Meeting with South Sydney Council, Saturday, 14 June, 1997). As noted earlier, a purpose-built development that houses students attending the nearby University of Technology, Sydney (UTS) had become the benchmark of undesirable 'tastelessness' for the area, which was reiterated during public meetings about the new proposal (Public Meeting with South Sydney Council, Saturday, 14 June, 1997). Heritage and taste were inextricably intertwined throughout the overt protest discourse. For RAG, acceptable development depended on acknowledgment of heritage authenticity (Lowenthal 1985). In this case, such authenticity was deemed tasteful. During the campaign, RAG negotiated with the SSCC and the developer over what would be 'acceptable', and what was 'unacceptable' style. As Jacobs (1992) has observed, some activists have access to certain aesthetic discourses that are better able to register in wider planning discourses than others. RAG certainly gained access to the decision-making process regarding this development and the most powerful members were eventually called upon to participate in the final planning stage. The experience of other concerned residents was that they became dependent on RAG to represent their concerns, and 'the community' (as represented by RAG) would be monitoring any new development. Acceptable development, according to RAG, had to acknowledge the existing built heritage of the area (with certain reservations). The second criterion of acceptability was that new developments required the input of RAG's expertise and taste.

RAG in action

RAG resident action started with the formation of a self-appointed 'committee' of local professional people who mobilized to combat a proposed conversion of an old factory site to apartments/condominiums. Fuelled by a concern about the loss of an old industrial building, the committee held meetings and reported to the wider community by letter-box-dropped newsletters. There was one 'street meeting' on 8 February 1997, where local residents were given various instructions about how to object to the development, and were also then invited to attend a public council meeting.

As part of its campaign, the RAG committee carried out a questionnaire-based survey of the Darlington area to assess the opinions of the local people about the proposed development. The survey provided several responses from which residents could choose. The results indicated that 35 per cent of those surveyed wanted only houses to be built (i.e. no apartments). Another 29 per cent wanted no development *at all*. A slightly smaller number, 26 per cent preferred that the warehouse be saved and the remainder of the site be developed. Only 10 per cent supported development of the entire site (RAG Newsletter #4, 24 February 1997).

The response tally indicated that 64 (35 +29) per cent of residents *did not want* the development of apartments on this site. In response, the RAG committee then met with councillors and planners from SSCC, and the developer. The RAG committee then submitted a new development proposal to SSCC. This proposal included a surveyor's report and new architectural plans *for a development of units* (apartments/condominiums). Instead of representing 'the (surveyed) community' by fighting the development, the self-elected RAG committee funded (with calls for donations from local residents) a redesign of the site based on their ideas about what would be the most appropriate development for the site.

The new design required that 'the façades, skyline and materials used ... [be] in keeping with the surrounding Conservation Area' (RAG newsletter #5, 3 March 1997) and the façade of the small warehouse was saved in the redesign (RAG newsletter #8, 7 July 1997). The developer capitulated to the RAG specifications and modified the development. The new version of up-market/up-scale apartments was designed not to attract students. This was then accepted by SSCC and the outcome was a boutique serviced-apartment/hotel complex.

The redesign of the site was promoted as a 'win' for local area resident activism, and local 'heritage' ('Thank You Darlington!' RAG newsletter #8, 7 July 1997). The spokespeople of RAG applauded their neighbours for gaining the right to have a say about heritage and development in their area. That 'say' was, however, limited to the options presented on behalf of the Darlington residents who were not part of the final decision making process. The notion of 'an alternative' plan *seemed* to provide an option for community-minded individuals.

The taste of good heritage

The façade of the small warehouse, as representative of the former diversity of land use in the area, was saved because of the RAG

campaign and the council's preference for 'recycling' over demolition of
old warehouses in the area.

> The existing warehouse [at 74 Ivy Street] although modest ... does repre-
> sent the industrial history of the area ... continued demolition of existing
> buildings within the area would ... diminish [the area's] heritage value.
> (SSCC Minute paper Item 9, Planning and Development Committee, Ref
> 2378/JGP/FMM, 11 June 1997)

Along with a campaign to save industrial heritage, the use of 'taste' was
a powerful force in the RAG campaign. The final decision about the
number of units and the colours and features of the new development
followed the RAG plan. The acceptance of this 'alternative' plan was
presented as the Council's compromise with the local community. The
question remains as to *who* was included in this version of local community. For one interviewee, RAG was not committed to human diversity.

> In the end, all they were fighting for was balconies that didn't overlook,
> overshadow their building, a little bit more open area on the corner ... and
> I said ... you've lost the plot. Save the buildings? They managed to save a
> few façades ... All they were after was a residential development that
> looked OK and didn't impact on them, but ... we need these warehouses,
> we need this sort of usage [artist and photographic studios ...]. (Interview,
> 29 April 1998)

Regardless of RAG's achievements and their failures to represent and
protect the (potential for) diversity, the process did invoke widespread
agreement on one issue: the need to preserve *built heritage* in the area.
As considered earlier in this chapter, this desire has prioritized the
symbols of select pasts, of terrace houses and old industrial façades,
over contemporary expressions of human diversity. Such preoccupations
have assisted in the disengagement from concern about current lived
diversity. It is simply easier, and more comforting, to remain preoccupied
with protecting the symbols of a more familiar past: in this case the
artefacts of the not-so-distant colonial past. I call this phenomenon
'architectures of denial'. By this, I mean that human diversity is denied
as differences are subsumed to an established register of heritage impulses that excludes all others. The deeply embedded desires to preserve
the reminders of colonial pasts protect the increasingly diversity-free
spaces established through the monocultures of gentrification. These
have become part of the escape from the everyday realities of a colonial
aftermath that has produced mass migrations and, in the extreme

example that is ever-present in downtown inner Sydney, Aboriginal poverty. The aftermath of colonial dispossession, which is so well represented just around the corner from all this important heritage, has become just too difficult to reconcile.

Architectures of Escape 1: Into the Past

In the neighbourhoods surrounding The Block, imaginaries of the past are being constructed to consolidate specific entitlements to place. In a contradictory way, they also envision escape from aspects of the everyday of that place. Such escapes are engaged by looking to romanticized and select pasts, and by protecting those memories. A second strategy of escape, of avoiding the everyday, is through the evocation of *elsewhere*. For those living in New York style loft apartments, the harsher realities of 'the street' can be distanced through new urban lifestyles (Shaw 2006)[41]. Both strategies of escape produce *indifferences* to the here and now, through denial or elision of pasts that have contributed to the current conditions, particularly with the expanding gap of inequity elucidated daily by the presence of The Block community. Through denial of the impacts of the colonizing encounters on Indigenous peoples, by privileging the whitewashed (Gabriel 1998) versions of history, connections with current dispossessions and poverty are also denied.

Georg Simmel (1995 [1903]) referred to the deployment of a 'protective organ', and in the environs of The Block, insulations are developed against the unpleasantries of urban life. Alan Latham (1999, following Sennett 1990, 1994) has defined indifference as 'the embodiment of a subtle economy of denial'. Indifference is perpetuated through embedding oneself, at least in part, in some aspect of a fantasy, of other times and/of other places. The activation of (and, at times, obsession with) accompanying routines provides distraction from the everyday. Engagement with turf protection, and the protection of fine architectural objects, is a useful distraction from other, more difficult political issues. Lifestyles are built up around such pastimes, and provide a comfort zone that detracts from (the pathologies) of the local. There is an imaginary of other times in inner Sydney, of vague and varying notions of *the good old days* that are indifferent to recalling those aspects of class and race relations that make the memories less palatable. Such imagined pasts, therefore, deny any repercussions of the past in the gentrifying present. Only the partially remembered pasts, glorified through built heritage and specific objects, are brought forward into the present, celebrated, and preserved. This renders all other local concerns, such as Aboriginal or other heritages of the inner city, as inconsequential. Heritage making legitimizes certain presences, in the present.

Conclusions

The detail in this chapter has added empirical 'flesh' to the proposal that the fantasies of exclusionary heritage designations, and the defence of space against Aboriginal activities, be they drug use or employment services, are processes of whiteness. Such processes have built on a notion of entitlement, through history, and have sought to arrest or deny the paradox of Indigenous settlement in the post-colonial experience of living in an increasingly historically significant city.

I have traced examples of the selectivity of heritage designations in inner Sydney and realized that, as Lowenthal (1985) has suggested, the past is indeed like a foreign country, with vestiges of Ye Old England reminding us of other days, and other places. 'Old stuff' has the capacity to invoke escapist fantasies, and the very selective designations of 'heritage', from chunks of Victoriana, to bits of tacky retro-chic, have special significance. Somewhat ironic, but highly strategic, is the exclusion of the significance of other histories, as amplified by the story of a return of Aboriginal people to the city of Sydney. The fantasies of escape conjured by 'funky old stuff' go beyond simple desires to recycle resources. These bits and bobs have the capacity to produce nostalgias, for the good old days when life was simpler, when we knew our place and accepted our lot. These were the days before all the troubles, the social upheavals of civil rights movements, and the rise of Indigenous politics in Australia in the 1960s. Those days have been remembered and glorified through artefacts of the heritages associated with 'Anglo' whiteness, be they tasteful or kitsch. But history continues to be truncated, and compartmentalized, as it unfolds through the production of very specific heritages. For Herzfeld (in Rapport 1995, 645), the sanctioned discourses of heritage: '...erect...a set of rigid, iconic, homogeneous, national categories which stereotypically define who belongs and who does not...'

These complexly produced categories are identifiable. In the new residential, postmodern context of a globalizing post-colonial city, the renewed interest in 'heritage', as (neo-)colonially encoded built formations, has served to embed a specific identity politics that can be traced as it is manifested in representations of, and *by* 'heritage'. In the case of inner Sydney, the built form has served a project of reinforcing and reproducing a heritage of 'whiteness'.

In a process of reinvention, a time-line of 'heritage' production, in the area surrounding The Block, started with Victorian terrace houses. It then expanded to include industrial landscapes and then opened up to the even more recent retro-chic and faux-heritages. Migrants, and their efforts to preserve and utilize crumbling housing stocks, the labours of the working classes, and the emergent 'black politics' of the 1970s, through the formation of The Block, remain external to the ever-evolving heritage imaginary, and its protection(ism). Preoccupation with heritage entitlements, which recently expanded to include pre-Aboriginal 'Nature', in the form of parklands or 'open spaces' for the former Wilson Bros. factory site, has proved to be a useful way to deny current realities of class relations and neo-colonialisms. Because Aboriginal people and associated places have been disengaged from the experiences of more recent pasts, they have become museum-like objects (Wasserman 1984, 1994, Thomas 1994), or disregarded or, as has occurred in inner Sydney, actively repressed through resident protectionism.

The desires to contain and ultimately arrest an urban Aboriginal presence in inner Sydney, have found expression through localized resident activism and government responses to issues such as drug use, which, in this part of the city, continues to be racialized. Responses, by residents and governments, have left discursive and non-discursive evidence, markers that have been traced to reveal the ways that whiteness has engaged in the empowering processes of negotiated space and heritage.

In the guise of what seems to be an increasingly pluralistic city, which is duly drained of politics (Kearns and Philo 1993), there is, in fact, a very specific exclusionary politics. The active remembering of specific versions of history and entitlement have become forms of escapism, ways to retreat from the realities of everyday life and the pathologies of poverty and dispossession, particularly of a racialized 'other'. Gentrification has brought with it a set of highly volatile and exclusionary resident politics that want to reclaim inner-city spaces for those who have gained entitlement to them through their connections with particular histories. These selective remembrances help to maintain territories that were gained in the past, and seem set to continue to be fought for in the future.

Notes

1   Efforts to 'redress' the absences of other stories from the mainstream account do exist, and one outcome has been the 'History Wars', and for a summary of these conflicts see Macintyre and Clark (2003).

2   This means that 'Council needs to consider the impact of the proposed
    development on the attributes and character of the area. This includes
    assessing the impact of both the demolition of...existing buildings and
    the relationship of...new development with the surrounding conservation
    area' (SSCC Minute Paper Item 9, 11 June 1997, 5, 6).

3   Database # 001785, File Number: 1/12/033/0011.

4   See chapters by Rubino on Brazilian cities, and Fujitsuka on Tokyo in
    Atkinson and Bridge (2005). For Chinese cities see Li L. and Soyez D.,
    Industrial (heritage) tourism in newly industrializing countries: status,
    barriers and potentials. Paper presented at the International Geographical
    Union 2006 Brisbane conference.

5   See Jackson (1999b) for review.

6   The Green Bans were a form of development 'black ban'. The Builders'
    Labourers Federation (BLF) was a strong trade union that used Green Bans
    to halt developments that were considered to be detrimental to the envir-
    onment, or threatened 'working-class' or heritage housing (Roddewigg
    1978, Mundey 1981, Jakubowicz 1984, Burgmann and Burgmann 1998
    and, for Aboriginal housing, Anderson 1993a, b; Burgmann 1993).

7   'Cultural cringe', in the Australian sense, refers to a belief that Australia is
    somewhat second-rate, culturally. It is a colonial legacy that perpetuates a
    deference to the superiority of Europe, and in particular, to England.

8   The main language group, according to Anderson (1998, 217), is Banjalang
    but there are also Eora (Sydney), Wiradjuri (Nowra), Kamilaroi and others
    from throughout NSW, and Queensland.

9   Beyond the scope of the current discussion are the following consider-
    ations: cheap inner urban property was/is often sought by groups who
    needed/wanted to be grouped 'together', such as migrant groups; gay men
    and lesbian women and other survivors of suburbia. The economic/cultural
    'mix' of such imperatives are contingent upon a number of factors, not least
    of which include issues of marginality and exclusion in the largely 'white'
    heteronormative suburban landscape. Ethnic concentrations, so specifically
    inner-urban in the past, now occur in suburbs but tend to clump (eg
    Cabramatta, Auburn, Chatswood) rather than 'assimilate'. In recent years
    communications technologies have enabled some unsettling of, for example,
    heteronormative dominance.

10 Students did, however, benefit from the glut in rental stock until the late
    1990s.

11 Residex figures for March 1998 were: median house price for postcode
    2008, Darlington and Chippendale, was $290,000 and for postcode 2021,
    Paddington, $580,000. These median readings span the range of house sizes.
    According to Real Estate Institute figures median house prices in the Dar-
    lington and Chippendale postcode area (2008) rose by 3.5 per cent between
    March 1998 and March 1999. This was in a market of general decline in
    house prices (for example Paddington house prices *dropped* by 5.3 per cent).

12   In mid 2002 an almost derelict mid-sized terrace house sold in Darlington for an unprecedented $700,000. The market determined that this house was more valuable than similar renovated houses. Also, Darlington had attained an advanced stage of gentrification regardless of the presence of The Block.

13   Some are only 3 metres wide.

14   According to South Sydney Council Local Environment Plan 1996, Heritage and Conservation, Darlington is a 'Conservation Area', with several 'Heritage Listed' buildings. Currently, this simply means that Council permission must be gained for alteration or demolition of a building (LEP p21). Victorian terrace houses have been demolished (e.g. on the corner of Ivy Street and Abercrombie Street, Darlington). Although The 'Darlington Conservation Area' is 'legally registered' with the Federal Government body, the Australian Heritage Commission, this has not protected it from local government jurisdiction. The Commission notes indicate that 'the data was mainly provided by the nominator and has not yet been revised by the Commission'.

15   'Chippo Politics' became the *Redfern Chippo Herald* when sponsorship shifted from the local branch of the Labor party to the Uniting Church. It is now the *South Sydney Herald*.

16   Examples include 'The Printery', 'The Piano Factory', 'The Cyclops' (once a popular brand of bicycle) and 'The Foundary'.

17   And on 15 February 2006, I accompanied a group from a North American university on a walk around the neighbourhood. An enthusiastic local resident joined us for a short time and reiterated this point firmly.

18   At the local council level a 'Heritage and Conservation' Local Environment Plan for South Sydney was introduced in 1996; At the NSW government level, the Heritage Council of NSW announced that a 'New State Heritage Register' for 'state icons' had been established (1998) by amendment to the 'Heritage Act' of 1977. At a national level the term 'National Estate' was adopted in 1972. There is the Heritage Commission, and the independent National Trust of Australia, established in 1950, which holds a classification of heritage register.

19   In the nineteenth century, a European scientist, Francis Galton, had graded the 'Australian Native' at 'at least one grade below the African...Negro race [which is]...two grades below our own' (cited in Sibley 1995). This 'level' of humanity, base and animal like, would naturally be caged in their zoo-like ghetto (see also Anderson 2000 for discourses of animality, and Dyer 1997 for Simian Irish Celt parallel).

20   On 22 March 2001, another resident and local business action group, the Chippendale Crime Control Committee, which had just formed, announced that it had hired a private security firm to patrol the streets near to The Block (*SMH*, 22 March 2001, 3). Most of the contributors to the fund were from the new apartment developments, hotels and businesses in the area just to the North of The Block.

21 I have used the pseudonym of Kommunity Action Kollective (KAK) to protect the identity of the members of a resident action group, many of who still live in the area.

22 The development had not gone ahead but the reasons for this are unclear.

23 It was later, quietly, re-established away from the media and (non-Aboriginal) resident spotlight.

24 The individuals are not named here, to protect their identity.

25 RRAG website at http://www.geocities.com/RainForest/Jungle/4021.html

26 at: http://www.geogcities.com/RainForest/Jungle/4021/policy.html

27 Vic Smith, Mayor of South Sydney Council; Clover Moore, independent state MP; Tanya Plibersek, Federal Labour MP.

28 According to Dear (1992, 291) AIDS related services have received responses from local residents varying from high acceptance to total hostility and strong resident action against facilities.

29 Resident action groups around inner Sydney have formed to fight the presence of existing welfare agencies in the gentrifying city.

30 Topics of concern included: 'honey-pot' effect of the health facility; normalization of drug use; lack of consultation; violence and crime; location of service (near housing); demonization of users and stereotyping; role of facility; racism; shortage of medical services; understanding need; inability to 'hear' and listen to locals; inappropriate consultation; consideration of the experiences/consequences of no needle exchange/facility. Stakeholders were listed as: Aboriginal Medical Service; Community [of The Block? Non-Block? Both?]; Residents [of The Block? Non-Block? Both?]; Police; Welfare groups; IV users and families; Local business holders; Schools and churches; Area Health Service; Needlestick sufferers; Future children of ex users; Children; Victims of drug-related crime; Redfern–Waterloo Development Authority; REDWatch; University students; Aboriginal Housing Corporation.

31 This is also 'The Redfern Aboriginal Coalition'. This mistake has come about because of confusion between The Redfern Aboriginal Coalition (RAC) and the Aboriginal Housing Corporation (AHC). The politics of the two could not be more different as one represents 'on the ground' issues (RAC) and the other (AHC) is the landlord organization. There have been tensions between the two organizations but these are beyond the scope of this book.

32 It is not clear who raised the concerns as South Sydney City Council lost many records during a severe hailstorm in April 1999, but it appears that the corrugated asbestos roof of the Wilson Brothers factory site was the concern. These are common all over Sydney. Many had to be replaced after the hailstorm (including the Council building) with corrugated iron, which is the cheap alternative.

33 In July 2000 drawings of the site proposal were posted on a wall of the Wilson Bros factory compound which showed that the very small terrace-house could be used as an 'artist in residence' studio.

34 At one of the later stage workshops in the community consultation process (8 May 1999), the appointed landscape architect stated that 'employment training schemes that produce profitability for other than local community' would not be allowed. Whether or not the RAC's usage was deemed appropriate or not for housing on the site is unclear. It is more likely that the high rents were prohibitive.

35 For comparison, I attended two of these duplicated meetings. They differed greatly even though the formal content was the same. One meeting was held in Holden Street, in the RAC building on The Block on Saturday 14 November 1998. This meeting was very poorly attended and chaired by a representative of the Redfern Residents for Reconciliation (RRR). The other meeting was at Darlington Activity Centre in Ivy Street, Darlington on Tuesday 17 November 1998. This meeting had a large, very vocal and at times racially charged crowd in attendance. A member of Redfern Residents for Wilson Bros., who gave a highly racialized and emotive presentation to the meeting, chaired the meeting.

36 I have used pseudonyms for LRAG and RWB (as with KAK).

37 Indigenous Australians have voiced their preference for the use of a capital letter if the term Aboriginal is to be used as a collective term for groups of Indigenous people. This usage has become convention in Australia.

38 Cf. Sennett 1976.

39 Sections 2.6 and 2.7 of the Draft Plan of Management are referenced to a facsimile, dated March 1999. The community consultation process began in November 1998, and finished in July 1999, which meant that the original purpose for the purchase of the Wilson Brothers factory site by the SSCC was 'on the record'.

40 RAG is also a pseudonym.

41 Also my unpublished PhD thesis: Shaw W.S. 2001. Ways of whiteness: negotiating settlement agendas in (post)colonial inner Sydney, University of Melbourne.

4

Cosmopolitan Metropolitanism (*Or* The Indifferent City)

Introduction

As the twentieth century drew to a close, so too did the end of the era of industrialization for many cities around the world, particularly those known as 'advanced' or 'western industrialized'. This closure was signified by a dramatic change to many inner-urban landscapes that once housed the heavy industries of a bygone era. With de-industrialization in some parts of the world, which meant the industrialization of others with the movement of manufacturing 'off-shore', defunct industrial landscapes needed a new purpose. Many of these 'brownfields'[1] have been redeveloped as housing. In this chapter I consider the conversion of swathes of former industrial landscapes into housing, using the city of Sydney to exemplify this wider trend. In inner Sydney, old industrial areas have been reborn as something special and new to accompany the city's emerging global status. The purpose of this chapter is to document processes of re-invention of districts, and understandings about new forms of housing and associated ways of living, through the (re)birth of the 'apartment'. In addition, I consider how such processes of reinvention have contributed to the rise of cultures of urban escapism; cultures that are imbued with whiteness. In the previous chapter, I discussed how the production of imaginaries of previous eras, evoked through preoccupations with heritage and its protection, have become mechanisms of avoidance – ways to disassociate from the pathologies of the present through preoccupation with (artefacts from) the past.

Another version of escape from the less pleasant aspects of city life has emerged with the building and re-imaging of apartments/condominiums with 'New York style'. Although this form of housing has links with the past, the capacity enabled for escape from the unpleasantnesses of the local is their particular speciality. The construction and promotion of fantasies of elsewhere, and the invocation of other times and places has produced new forms of urbanism. The promise of a Manhattan-like urban lifestyle offered with Sydney's latest housing form, carries a certain caché and offers sometimes amusing Manhattan-referenced nomenclature, clever building design, and high-tech surveillance technologies to 'transport' the consumer (metaphorically speaking) away from the necessity to engage with the unruliness of the local, particularly around the Redfern area. Following on from the previous chapter on the constructions of 'heritage' and heritage protectionism, and in particular the rise of enclave consciousness and malevolent resident activism, this chapter explores contemporary constructions of urbanism that have emerged with Sydney's 'Manhattanization', and their contributions to urban whiteness.

A cosmopolitan metropolis of whiteness

While Sydney embraces its emergent global status, the Indigenous community of The Block has become the city's 'black ghetto' in contemporary understandings, which has built on the idea first proposed in the 1970s as the basis of objection to its formation. Most non-Aboriginal people find this urban 'ghetto' image far removed from what Aboriginal people are 'meant' to be: traditionally tribal and nomadic (Thomas 1994). Non-Aboriginal residents of Darlington/Redfern, and others (such as the media, see Box 4.1), constantly feed, and are fed by what I have referred to elsewhere as the 'Harlemization' of The Block (Shaw 2000).

As portrayed in film, music, television programs and other forms of popular culture, Harlem has a dubious reputation as the quintessential ghetto of popular imaginings, regardless of the contemporary realities of its gentrification (Zukin 1995, Smith 1996). Obviously, there are striking differences between The Block and Harlem. Their histories, circumstances of ownership, external ties, political struggles, and symbolisms differ substantially. But the media's apparent preoccupation has usefully aligned their similarities through an image of the pathology of racialized disadvantage. The Block's continual reinvention as Sydney's own

---

Box 4.1   Ghetto-Referenced headlines

'Redfern "ghetto" – a BBC view', *Sydney Daily Telegraph*, 1 September 1983

'Violence: moderate blacks strive to get rid of ghetto image', *Sun Herald*, 15 December 1985

'Govt blamed for 'ghetto"', *The Australian*, 23 December 1985

'The two faces of Eveleigh St, Sydney's no-go area', *Sydney Morning Herald*, 25 May 1989

'Ghetto of anguish: As Harlem is for black Americans, Redfern defines urban Aboriginal life', *The Age,* 1 February 1997, 21

'ATSIC seeks $6m to demolish ghetto', *The Australian*, 4 February 1997, 3

'End of a ghetto/Bulldozers ready to raze The Block', *Daily Telegraph*, 29 March 1997, 15

'Block Out' (in which The Block is referred to as a 'sliver of South Bronx'), *Australian Magazine*, 14–15 June 1997, 13

'Redfern – a bogus war on drugs' . . . 'New York's South Bronx', *The Australian*, 1 February 1999, 13

---

inevitable 'black ghetto' is also part of the evolution of Sydney into a big international city; after all, big international cities have (racialized) ghettos. This image of a global cosmopolitan metropolis complete with its own 'problem area' is, however, generally external to the marketing of the sanitized version of New York style 'loft living' that I will turn to later in this chapter.

As Sydney began to experience its various forms of Manhattanism in the late 1990s, the former mayor of New York City, Mayor Guilliani, was busy introducing 'zero tolerance' measures to reduce the presence of unwanted 'others' from the reinvented Manhattan. Far removed from the origin of the term 'cosmopolitan', defined (in part) as 'free[dom] from national limitations or prejudices' (Thompson 1995, 302), the removal of homeless people and others from public places was part of a 'cleaning up' process. Ridding Manhattan of its 'black ghetto' was part of this cleansing (Smith 1996) and the island of Manhattan experienced 'whitening', as well as a thorough brightening, through gentrification.

In his study of whiteness, Ghassan Hage (1993, 63) proposed the existence of the 'cosmopolite', which he defined as an essentially 'mega-urban' figure . . . a class (and classy) figure *and* a western (or culturally and *physically* Westernised) person, capable of consuming 'high-quality' commodities and cultures, including 'ethnic' cultures.

For Hage, cosmopolitanism involved consumption of, rather than participation in, 'multiculturalism'.

Multiculturalism, at the centre of cosmopolitanism, often incorrectly includes Indigeneity. As some inner Sydney residents acknowledged (during research interviews), The Block community is part of the diversity of the city. One resident put it this way, 'they are entitled to have their bit of ethnic Sydney' (September 1997). The difference is that The Block is an Indigenous, rather than migrant community. It has arrived without a range of exotic 'ethnic' offerings. Urban Aboriginal people generally do not live 'traditionally'; they have not brought interesting objects/artefacts or ethnic Aboriginal foods to the city – there is simply nothing on offer for cosmopolite tastes (cf Thomas, 1994, Hage, 1997).

Without such obvious Aboriginal 'culture', the Harlemization of The Block has filled the gap, and provided a different cultural anchor for identification. The generic 'black ghetto' is highly recognizable. It is simply too easy to consider the urban Aboriginal community in Sydney in a similar way. 'Black ghettos' are, for many urban dwellers, no-go zones. They are places that have failed to fully integrate into 'ordinary' urban space, and follow the expected norms of those cities. And 'race' is the common emphasis. In the Sydney context, the ghetto label also renders The Block as foreign to Australia. With its de-Indigenized status, it appears to be even *more* inappropriate in an Australian city.

Meanwhile, the new lifestyles offered with apartment/condominium (or, as they are often labelled 'loft') living are very different to life in the suburbs, where the majority of Australians live. Cosmopolitanism (or, more specifically cosmo-multiculturalism) according to Hage (1998), is a form of 'European-ness' (or generic western-ness), and relies on the fantasy of an ethnically invisible hegemony. For Sydney, Manhattan(ism) and apartment living, have reached a pinnacle in contemporary urbaneness. Many now flock to the 'glamour' of civilized (western-style) high-rise living in the city. The sounds of construction that come with rampant redevelopment (of existing properties, and new apartment blocks), are also accompanied by the distinct and familiar cries of police helicopters overhead, and sirens in the street. Alongside the clean and safe packages of apartment living, of caricatured ('white') Manhattan, lurk the hard realities of increasingly polarized urban lives.

This chapter documents the rise of Sydney's 'Manhattanization', which is the story of a city's vertical evolution, and the 'dormitorization'[2] of former industrial and commercial areas. The second part of this chapter considers another theme of urban escapism, the rise of 'New York style apartment' landscapes, and the associated strategies for living and coping with the challenges of the progressive postmodern city of the twenty-first century.

Ways of seeing the city

> City imaginings are increasingly detached from urban reality...promotional rhetoric and issues of community conceal, and partly effect, gross economic and political inequalities...the inner city landscapes of gentrification and spectacular consumption divert attention from, and partly produce racialised poverty. (Goss 1997, 181)

Cities are 'systems of communications [that] tell...us who has power and how it is wielded' (Short 1996, 390) and to make sense of these communications, cities can be 'read' like poetry. As cities are reshaped, and parts are rebuilt, new meanings are created and projected. In geography, there has been 'an increased recognition of symbolic and representational realms (figurative and discursive)...in the constitution and mediation of social and material processes' (Jacobs 1993, 829). As with other texts, messages 'hidden' within architecture can be read. Michael Jager's (1986) study of the aesthetics of restoration of Victorian terrace houses in the gentrifying inner-city neighbourhoods of Melbourne was a landmark reading of architectural representation. Darrel Crilley (1993) has also carefully observed the use of architecture to convey meaning in the marketing of new housing developments in London and New York.

Pierre Bourdieu's (1984) notion of cultural, or symbolic, capital has provided an elegant theoretical framework for making sense of the meanings within cultural formations, such as architecture. He considered the 'social uses of art and culture' and the way that taste functions as a marker of class (Bourdieu in Jackson 1991b, 220). In other words, Bourdieu identified the power of meanings within such cultural formations, to convey markers of class. Drawing on Bourdieu, Jager (1986) identified the emergence of class distinction with gentrification, and the preservation and emphasis of specific aesthetic features within the 'heritage' landscapes of inner Melbourne. Such distinctions can also be created through clever marketing, and in the case of Sydney's burgeoning apartment landscapes, taste and class have become powerful markers of distinction for those with the (economic) capacity to discern the buying possibilities in the first place, and then to choose from the variety on offer.

As noted in the previous chapter, Crilley (1993, 231) identified a shift in research to considering the 'images and representations *of* development rather than to the images projected *by* development'. He remarked on the use of heritage motifs in new developments and their capacity to attract nostalgic interest. These historically referenced motifs 'imbue [d]...[a]...neighbourhood with the instant aura of tradition and familiarity...[that] tap into the gentrification market...by catering to the middle-class tastes there being expressed' (Crilley 1993, 233). In this sense, the architectural postmodern turn, which included the addition of decorative heritage-referents, has been a very clever marketing strategy. The shift from modernist to postmodern architectural forms has enabled the harvest of the symbolic capital of heritage, and produced cultural codes. As Crilley (1993, 233) astutely observed, 'under the rubric of post-modernism, architecture...had primarily become a matter of the systematic, purposive manufacture and marketing of commercialised meanings'. Rather than provide a reading of the consumption of such motifs, Crilley raised the need to consider the *production* of meaning. So, what meanings do new developments convey, and what is the purpose of doing this? Crilley's study of new building developments revealed a deliberate encoding of a pop-culture version of 'heritage'. One purpose of such encoding was, obviously, to sell real estate but, as he pointed out, there was more to this.

> Buildings themselves are designed to 'read' as gigantic outdoor advertisements...nostalgic architectural compositions establish...the cultural theme and ambience conducive to up-market consumption...The imagery of architecture seek[s] to persuade the public of the virtues and propriety of the property capital commissioning it. (Crilley 1993, 236, 237)

The specificities of inclusion, or exclusion, promoted with such encoding are, therefore, also identifiable. In Chapter 3, I mentioned that Crilley is suspicious of the notion of 'diversity', which often means the provision of variety in housing, or architectural styles on offer, in new developments. He identified how the term 'diversity' was used to sell to very specific and narrow markets, the well-heeled 'middle classes' in Crilley's

US and UK examples, rather than to include a range of different people. These developments did not provide genuinely affordable housing for a lower-income market, nor did they cater, for example, for extended families.

Large-scale investment in the redevelopment of inner cities is, therefore, about much more than the material rebuilding of depleted areas. Linkages occur between urban transformations and formations of global capital, from corporate headquarters to fast food outlets and 'fashion streets'. New urban lifestyles have also become well-recognized components of the cultural capitals of transforming neighbourhoods. But, to avoid proliferating what David Ley (2004, 151) has referred to as 'globe talk', which is a 'master discourse reproduced in intellectual, policy and corporate circles' that privileges economistic explanations for urban processes, in what follows I also engage with contemporary understandings of 'urbanism', or ways of living in cities, within the context of the specificity of (a) place, and place–base identifications. To activate these contemporary understanding of urbanism, I have found it useful to be mindful of ideas posited by urbanists of yore, of Louis Wirth's (1995 [1938]) urbanism 'as a way of life', Lewis Mumford's (1995 [1938]) exploration of the 'culture of cities', and in particular, Georg Simmel's (1995 [1903]) musings on the 'mental life' of urban dwellers, their fears and responses. Acknowledging, and drawing on such understandings of urban life has the potential to enrich cultural geographies of cities. So, I turn now to Sydney's recent phase of urban transformation, which has been widely identified as its 'Manhattanization'. The wholesale transformation of parts of the Central Business District (CBD) and former industrial areas into apartment blocks has invoked new and increasingly exclusive ways of living in the city.

The most recent manifestation of Manhattanization, the advent of 'New York style apartments', began in the late 1990s. As gentrification cycles have matured, and existing housing and land stocks have escalated in price, property developers turned to the somewhat undervalued, and often under-utilized landscapes of old warehouses, small factories, former breweries and flour mills/silos around the CBD. With the rebuilding of these former industrial landscapes, the consequent apartment boom[3] heralded a dramatic departure from the usual taste cartographies for Australian housing. To sell the hitherto suburban nightmare – the earlier versions of apartments were dubbed 'flats' or 'home units' – required complete reinvention. The result is that apartment living in Australian cities is now widely associated with urbanity and global cosmopolitanism. The conversion of old buildings into what have been widely dubbed as 'loft' apartments[4] was followed by an avalanche of new loft-apartment construction. The advent of Sydney's own version(s) of that now highly recognizable global trend, which referenced the Manhattan phenomenon of 'loft living' (Zukin 1982), signified a massive reorientation from suburban to urban living in Australian cities.

In her account of a similar emergence of loft-apartment developments in Montreal, Julie Podmore (1998) identified what she called 'SoHo Syndrome' as:

> a spatial and cultural process that involves more than simply copying the aesthetic of SoHo [New York] as a redevelopment strategy...cities are 'locales'...[SoHo Syndrome is] more than a universal valorization strategy: it is a socio-cultural process that involves a complex web of relationships between place, identity and the media, that is diffused to, and (re)produced in, divergent inner city locations. (Podmore 1998, 284)

There are complex socio-cultural processes at work in what I have identified as Sydney's SoHo Syndrome (Shaw 2006). In this chapter, I ponder the extent to which the production of cultural attributes, which have been assigned to the consumption of the new residential landscapes, have produced new fields of whiteness in inner Sydney. The previous chapter mapped the invention of heritage(s) and associated protective activities that have privileged and consolidated whiteness through the denial of all pasts other than those that reference colonial settlement and its inheritances. In this chapter I present a residential landscape that has been purpose-packaged to appeal to a specific type of Sydney property purchaser: those with the economic capacity, and proclivity, to opt for an inner-city apartment over the more traditional terraced house, or a house in the suburbs. I posit that the specificities of inner Sydney as a locale, and the complexity of issues of identity that have been promoted (and *not promoted*) by the media and real-estate marketeers, are as pertinent to Sydney's version of SoHo Syndrome as any other, more external, global forces.

The commodification of lifestyles, particularly those associated with *urbane* style, has gained unprecedented commercial importance in the selling of inner Sydney apartments. For those interested in protecting heritage, particularly with the rise of interest in the preservation of industrial heritage buildings, Sydney's 'New York style apartment' landscapes include apartments in recycled or part-recycled warehouses.

Where there was no old building to recycle, specific urban lifestyles have been offered to promote the sale of new-build apartments. The promotion of the new urban lifestyles, or urbanisms, has been the basis of a persuasive marketing campaign with all forms of 'New York style apartment' living in Sydney, be they in recycled, part-recycled or new buildings. The specific marketing strategy that was used *en masse* to promote Sydney's 'New York style apartments' began in the late 1990s. To market the new way of living, fantasies were created based on the image of ('white', up-market/up-scale) Manhattan, and included the promise of fortressing technologies that were not dissimilar to those found in gated communities (Davis 1990, Hillier and McManus 1994, McKenzie 1993, Morgan 1994, Soja 1990). Elegant Manhattan-like living that specifically referenced the SoHo neighbourhood of Manhattan promised that inside the homescape, the highly security-conscious apartment with the car safely locked in the basement, lifestyles of elegance would proceed unhindered by life on the street. As Jager and Crilley have both shown, in examples from other cities, the production of specific property types can also produce specific, and orchestrated, meanings. In the following, I reveal that counter to the logics of property marketing for 'New York style apartments' that appeared to bring New Yorkness *to* the city of Sydney, other Manhattanisms have taken place in this city, and been socially institutionalized (Collins 1995), well before the latest apartment boom. Historical and contemporary promotions of the city have revealed its sometimes obvious and at other times subliminal identification with the metropolis of a very 'white' version of New York.

Manhattan Dreaming (in Sydney Australia)

> Manhattanism ... [is] an unformulated theory ... whose program [is] to exist in a world totally fabricated by man [sic], ie to live inside fantasy ... Manhattanism ... [is] an explicit doctrine that can transcend the island of its origins to claim its place among contemporary urbanisms. (Rem Koolhaas 1978, 7)

The gentrification of inner Sydney and associated escalation of property prices meant that buyers and renters had eventually to turn to gentrification's 'final frontiers' (Smith 1996), the neighbourhoods of Darlington/Redfern and Chippendale that surround the Aboriginal settlement of The Block. The influx of population to the inner city has dramatically urbanized the lifestyles of many Australians, and of others who have moved into apartments from abroad. Training a keen eye on this residential refocus on the inner city and Central Business District of Sydney, and wanting to encapsulate the shift away from conventional ideas about housing, the media and others loosely engaged the term 'Manhattanization'. This term became embedded within the lexicon of urban living in Sydney, in the late 1990s. As with many other cities engaged in meeting the demands of economic globalization and employment trends (Sassen 1991), inner Sydney began to transform. Following on from 'industrial chic', 'New York style apartments' flooded the real estate market. By the late 1990s, Sydneysiders had embraced the term, and the housing form, known as *apartment*. But, contrary to popular understandings, Sydney's identification with Manhattan actually began in the early part of the twentieth century. In its ongoing search for a global identity, city promoters looked to larger and older cities for inspiration and direction. The implications of this ongoing fixation are considered in more detail, in the next section.

The sudden flurry of promotions that promised Manhattan-like lifestyles in Sydney, in the late 1990s, was not, however, unique to this city (on New Jersey, see Cole 1987; Minneapolis-Saint Paul, Jackson 1995; London, Bowman 1995; Montreal, Podmore 1998). In her study of Montreal, and in line with recent gentrification research, Podmore (1998) identified that most investigations of 'lofts' had tended to emphasize their consumption, as (investment) commodities, rather than the production of the built form and lifestyle component, and the meanings encoded within. For example, Jackson (1995), had identified the consumption of lofts through arts re-investment, and heritage preservation, and Cole's analysis (1987) utilized Zukin's (1982) 'role of artists in gentrification' thesis. Bowman (1995) suggested that the preservation of warehouses in Hobart (Tasmania) occurred as a direct result of occupation by artists and community activists.

There is evidence to suggest that the Sydney version of SoHo Syndrome has transcended the influences of trailblazing arts reinvestment or even desires to preserve industrial heritage. I propose that other localized relationships have provided the cultural capital and some material foundations for the rise of apartment, or 'loft', development. As Podmore (1998) suggested, to fully understand 'SoHo Syndrome', spatial relations, aesthetic descriptions and social locations also require interrogation.

In this chapter, I want to highlight the importance of localized specificities that may *appear* to have similarities to the trends identified in other cities in the production of the 'New York style apartment' – which also *seems* to be a generic housing form. In the case of Sydney's SoHo Syndrome, I have found a reserve of unacknowledged but traceable Manhattan imaginaries that have contributed to the latest formations of housing and urban cultures. These localized socio-cultural moments have already whispered of Sydney's New York-ness, and such moments have been embedded, over time. The kernels of cultural capital that follow have been called upon in the legitimization of Sydney's current manifestation of SoHo Syndrome. And yet, the recent spate of promotions for 'New York style apartments' has not, in any tangible way, acknowledged local histories of this part of their production. Instead, recent and current promotions have relied on fantasies of Manhattan and, in particular the concept of 'lofts' and urbane ways of living associated with this housing form. For this latest manifestation of SoHo Syndrome, both property and lifestyles have had to sell and this commodification has relied, somewhat deliberately, on bypassing local geographies, and histories. So a task of this chapter is to acknowledge a history of illegitimate warehouse/loft occupation that hitherto remained largely unrecorded (Shaw 2006). Revealing such a history explodes the myth perpetuated in marketing: that 'loft living', in Sydney, was derived solely from Manhattan's SoHo district.

In the next section, a history of Sydney's Manhattanization is outlined. Then, I contextualize a very specific discursive environment that has dominated the recent phase of urban restructuring. A process of denying the local and its histories unfolds. So too does the promotion of an exclusionary lifestyle fantasy that has consolidated imaginaries of whiteness, in inner Sydney. However, unlike the promotional materials that deny the local, I will show that Manhattan fantasies have not necessarily precluded the local – its pasts, nor its present. The latter part of the chapter considers how The Block's Harlemization has not only transcended the promotional discourses of Sydney's SoHo Syndrome, but consolidated its status as a 'black ghetto', and provided another form of cultural capital in the new and increasingly polarized residential city.

Projecting Manhattan

Prior to the recent emergence of popular interest in industrial heritage and metropolitan living, successive Australian governments had been busy presenting an image of the city of Sydney to the world. Since the

1930s, New York/Manhattan themes have been used in strategies to promote Sydney internationally. Rosaleen Smyth (1998) documented a century of government propaganda films that were 'designed to attract immigrants, tourists and trade, and win national prestige' (Smyth 1998, 2). She identified that, 'at the end of the 20th century Sydney is a multiracial metropolis . . . thanks to a Manhattan-like skyline . . . Sydney has emerged as a 'highly imageable' city.'

Smyth did not openly acknowledge the underlying projection of Manhattan-ness that is in evidence in her account. However, hints and depictions of Sydney's New York-ness pepper the narratives and images, and an evolving New York/Manhattan metropolitanism can be 'read' throughout Smyth's record. For Harvey (1989), and others (Kearns and Philo 1993), the promotion of cities represents the 'postmodern hallmark . . . of place marketing' (Kearns and Philo 1993, 30). Smyth has provided a history of film-making by successive Australian governments that used a pastiche of images, and other triggers of Manhattan, to promote an urbane metropolis.

From Smyth's account, the first promotional film was *Sydney's Harbour Bridge* (1932). The film's commentary stated that: 'The north shore [of Sydney] would now be able to enter into its destiny as "the Brooklyn of the south seas"' (Smyth, 1998, 5). Brooklyn, in New York City, lies just across the famous bridge (of the same name) from Manhattan. Sydney's North Shore had been similarly linked to the Central Business District, the city's hub, to become Sydney's version of Brooklyn. Images of Manhattan were also presented in films such as *A Nation is Built* (1937). A promotional poster for the film exhibited 'images of the first settlement [in Australia] . . . backed by a cityscape of tall buildings' (Smyth 1998, 6). Promoting Sydney's maturing cityhood, with a timely silhouette of tall buildings, this film was screened at the 1939 World Fair in New York. Also in 1939, *Australia is Like This* emphasized:

> How like America Australia and, particularly, Sydney is . . . 'The first thing that hit us was that Sydney was so much like a city back home. This building might be in Chicago, New York or San Francisco while the skyline could be Manhattan', said one of the GIs. (Smyth 1998, 7)

By 1947 the mood had shifted, slightly, to a more generic form of 'metropolitanism', with an emphasis on sophisticated urbanity. *City In the Sun* set out to 'document the mood of Australian metropolitan life' and aimed to assure prospective immigrants that Australian cities could 'boast all the sophistication and conveniences of city life in older countries' (Smyth 1998, 9). Continuing the theme, *Saga of a City* (1957) described Sydney as 'the metropolis of the South Pacific' (Smyth 1998, 10) and, as Smyth pointed out, Sydney was depicted as a sophisticated and modern centre, on the 'periphery' of the non-European world. Moreover, in *Saga of a City*, Martin Place 'the heart of the city' was presented as 'Australia's Champs-Elysées, Trafalgar Square and Fifth Avenue' (Smyth 1998, 10).

By 1965, in *City of Millions*, Sydney's skyline had changed and 'the stock exchange . . . [was] presented as the pulse of the city's commercial life' (Smyth 1998, 11). In many later promotional films the themes of sophistication, international finance, urbanity and metropolitanism were repeated. Sydney was promoted as modelled on the British Empire's *first* great 'white city'. For Smyth (1998), Sydney was the empire's 'second greatest white city' (from the paper's title *From the empire's 'second greatest white city' to multicultural metropolis: The marketing of Sydney on film in the 20th century*) but she did not, however, identity the first 'greatest white city'. Presumably, she was referring to London but the images that lurk throughout these filmic promotions of Sydney nodded towards New York.

Regardless of Smyth's intention with the paper's title, the direct and indirect Manhattan references documented in these promotions of Sydney served to advance an ethos of modernization and progress. For a swiftly suburbanizing city, striving for a post-colonization identity, such official promotions of its urban core served to align it with an image of driving progress in the 'New World'. From this perspective, the smaller and newer city could look to modernity, as well as mimic the antiquity of empire. The propulsion *forward*, towards modernity, did not deny Sydney's British heritage, rather it added another dimension to the existing caché of specific heritages.

More recent representations of Australian-ness, of larrikin masculinity and rugged or quirky individualism, presented in many Australian commercial films,[5] were not part of the filmic representations of past governments, with their emphases on Sydney's urbanity. These urbane images were part of the city's 'cultural encoding' (Jackson 1995, 181) that was embedded over decades, during a city's formative years. Another layer in the accretion of encoding Sydney's affinity with New York has occurred through the built environment.

Loft living 1: warehouse occupations

In a newspaper article published in 2001, the (then) president of the Royal Australian Institute of Architects, Graham Jahn, identified two distinct stages in Sydney's apartment culture:

> the first was in the 1920s to 1940s period, which was based on a New York-style apartment...[and the second is] a more sophisticated range of developments providing an alternative to living in...houses. (*Sydney Morning Herald*, 3 February 2001, 4A)

Between these two events on the official record of 'apartment culture', I would like to add a transitional stage – of warehouse occupation – that I propose was highly relevant to the phase of 'sophisticated...developments', *aka* 'New York style apartments'.

The first application to convert an old building into apartments was made to Sydney City Council in 1979 (Roseth 1981) and marked a turning point in the history of Sydney's residential landscape. In 1981, the head researcher at the former NSW Department of Environment and Planning, John Roseth, published a short article titled 'Residential conversions in Sydney' in which he reported on a new trend in 'recycling old buildings'. At that time, apartment/condominium development had begun to reshape the urban fabric of many cities around the world, and in Australian inner cities, former industrial and commercial landscapes were also being used for housing but often in less formal ways. Prior to the gentrification of the inner city of Sydney, which began in the 1960s, most people lived in suburban bungalows. With the march of gentrification during the 1970s and 1980s, and the restoration of Victorian houses that dominated the inner city, a glut of inner-city warehouses and empty CBD buildings provided another option for cheap housing. The city beckoned those who preferred the social interaction afforded by living amongst 'like-minded' people. Universities, colleges, diverse lifestyles and nightlife were all located close together. Some, oppressed by suburban life (Murphy and Watson 1997, Costello and Hodge 1999), found greater freedom in the inner city. The move from a conventional suburban house to a spacious but shabby warehouse/office building was, for those involved, a move to desirable urban chic. An oversupply of city buildings provided a vein of alternative accommodation throughout the 1970s and 1980s. Global economic trends combined with socio-cultural shifts to produce a quite lengthy moment of low-income accommodation within which bohemian lifestyles could flourish. Although not part of the recent spate of apartment promotions, warehouse occupations did help to preserve many buildings that now carry the 'loft' tag. More significantly, they promoted an appreciation of industrial chic in Sydney.

The rise of warehouse cultures

After World War II, residential accommodation in Sydney was in short supply and, as the adage goes, *necessity is the mother of invention*. From the late 1940s, the CBD had become home to the Sydney libertarians (later known as 'the Sydney Push'). This 'notorious group of anarchic intellectuals' (Coombs 1996, viii) deliberately shunned the suburbs and moved instead into low-rent boarding houses, run-down terrace houses, pubs and other less-desirable forms of housing. By the late 1970s, living in old, otherwise abandoned warehouses and office buildings had become an established lifestyle alternative all over the inner city and CBD. Because living in a warehouse or office building was 'illegal', as commercial or industrial zoning made no provision for habitation, the formal conversion process was slow. Most warehouses and old office buildings occupied business or industrial, rather than residential zones.[6] During the 1970s and 1980s, however, vacancy rates were high and landlords turned a 'blind eye' to residential occupation of their empty buildings. The illegality of leasing for purposes other than 'commercial' was of secondary importance to much needed rental returns. As Roseth (1981, 77) noted 'owners of the old office buildings [had]...few options'.

Many firms had decentralized and moved out of the inner-city areas to cheaper locations on the city fringes. Others had moved 'offshore', or succumbed to larger global economic shifts. For example, local clothing and footwear manufacturers had lost business because of cheaper imports from the 'Newly Industrialized Countries' of Latin America and Asia (Sassen-Koob 1985). Economic shifts, and the associated rise in availability of commercial properties in the inner city, enabled a variety of uses for large, increasingly unoccupied buildings until the enforcement of usage regulations that occurred with another turn in economic fortunes in the 1990s.

Roseth attempted several forecasts about the development of warehouse apartments but he remained somewhat oblivious to the displacement of low-income residents. He argued that because the conversion of existing buildings was cheaper than redevelopment, the saving would stimulate lower-cost purchasing power in the inner city. Unfortunately, his forecast was a little naïve but he did note that:

> Beyond these [economic] reasons, there hovers that rather nebulous motive – what society considers chic or fashionable – which does not always spring from compelling logic, but which seems to exercise a compelling force on behaviour. (Roseth 1981, 77)

Apart from acknowledging cultural considerations, and their apparent lack of logic, Roseth (1981, 77) did not completely ignore the political economy of gentrification either: 'certainly history has shown that when a residential area suddenly becomes desirable, the wealthy tend to displace the poor who used to live there.' He qualified this with:

> Residential development in the Central Business District cannot be seen in the same terms as the resurgence of those inner suburbs... [because] the Central Business District has been without a residential population, there is no question of displacing low-income residents... buildings proposed for recycling are usually those which have stood empty or *half-occupied* for some years. (Roseth 1981, 77, *emphasis* added)

The reference to 'half-occupations' is an interesting one. A first point is that although economic forces may be oblivious to the fate of people living in the margins and interstices of (mainstream) societal norms – and in this case, warehouse occupations were largely invisible – the interesting contradiction is that 'free market forces' actually provided the moment for such alternative housing opportunities. The culture of warehouse chic emerged to fill in an economic gap that had emerged with high vacancy rates. A second point about these 'half-occupations' is the invisibility of this informal sector, in official terms. Certainly, commercial or industrial leases did not record 'illegal' residential occupations, but it is the ambiguous half occupations that suggests that Roseth did notice *something*, though his reference remains an ambiguous one. Despite this lack of official recognition, however, warehouse and loft occupations in Sydney (and other parts of the world) were definitely noticed by those who were there. I now turn to the remembrances of the Sydney experience of warehouse living, and its entrenchment within the cultural foundation of the city.

Remembering warehouse occupations

Anyone who lived in inner Sydney (as I did) during the halcyon days of warehouse occupations remembers various warehouse communities, and how different they were to the predominantly suburban way of life at those times. Some warehouses gained folkloric status[7] and others were simply notorious. Some of the larger warehouses hosted huge dance parties during the 1980s, which were the precursors to the commercial dance scene, such as the internationally acclaimed Lesbian and Gay Mardi Gras (post-parade) party, held annually. An old department store, the Anthony Horderns building, sat on a site now occupied by the multi-purpose commercial and residential conglomeration called World Square, in the CBD. The Anthony Hordens building was home to artists, musicians and students, from the late 1970s to the mid-1980s. The Xerox building (Alexandria/Zetland, just south of the CBD) provided similar housing for the artistic 'fringe', as did various smaller buildings dotted around Darlinghurst/East Sydney, Surry Hills and Woolloomooloo which were all within a few kilometres of the CBD. Some warehouse occupations lasted until conversion of the buildings to apartments. These included the Spice Traders building in Surry Hills, which was converted to apartments in 1995, and the Shepherd and Newman Printery in East Sydney, which succumbed to redevelopment in 1997. The site of one of the most longstanding occupations, the Silknit Building in Surry Hills, was converted to apartments and up-market, 'arty' commercial spaces in 2000.

Although many buildings earmarked for redevelopment were demolished or partially demolished, artefacts of these cultural moments remain as objects, and memories. An example is the now demolished Slaughterhouse (a former abattoir and meat-packing building) in Redfern. The Slaughterhouse was a huge, constantly evolving multi-purpose concern with artist lofts, galleries, and recording and practice studios for musicians, as well as separate 'living spaces'. Renamed 'Renwick Street', when the old title had limited charm, this unofficial occupation lasted for well over a decade and left behind pieces of art, and recordings.[8] Other examples of residentially occupied warehouses with notable warehouse cultures included Alpha House, and the former Beta House in Newtown, south of the CBD. A property developer converted Beta House into 29 'loft apartments' and sold them off at 'luxury' prices in the mid-1990s. This prospect was not without contest, and the former occupants of Alpha and Beta Houses, and other locals, objected to development proposals for both sites. The local council received 30 letters of objection to the proposed loss of a viable art and music community.[9] The larger and more imposing building of the two, Alpha House, formerly home to a thriving rent-paying and at times infamous party-throwing community, sat empty and gutted throughout the 1990s. The developer submitted a development application in 1993, ready for the impending boom in apartment developments, but the conversion stalled. Alpha House sat empty for many years as a testimony to developer speculation and, perhaps less obviously, to former informal occupancy[10]. A CD-Rom publication immortalized its former life, as did a Sydney newspaper in an article titled 'Home is where the art is' (*Sun Herald*, 15 August 1982, 18). After running a story on the 'former glory' of Alpha House, a rash of callers – those who had lived there or knew about it and wanted to know more – telephoned a local independent newspaper, the *Newtown Times* (Pers. Comm. *Newtown Times*). Some current residents remember this kind of inner-urban underground culture, and one terrace house occupant in Darlington noted:

> There was light industry then it became a lot more artistic in usage. Those sorts of people...need large warehouse spaces close to the city...they're getting squeezed out, there's nowhere for them to go, and so a whole artistic community in Sydney is just being pushed out. Where are they going to end up? I mean we've really got the death throes on a lot of artists here. (Interview, 1998)

Although several warehouse communities lived on, the cheap housing rental alternative overall came to a halt with a sudden increase in real-estate prices during the mid to late 1980s. At that time, local governments deployed policies to encourage building 'recycling'. Roseth (1981, 77) noted that

> ordinance 70 allow[ed] councils full discretion regarding the extent to which the requirements it would make of a new building...applied to the alteration of an old one...to convert to residential.

This meant that once economic factors were in place, existing warehouses and old office buildings became attractive to developers and the conversion phenomenon. Some buildings were converted into warehouse apartments by the late 1980s but the economic recession of the early 1990s stalled the conversion process. Many developers sat out the recession, as did many old buildings, until the mid- to late 1990s when Sydney's own SoHo Syndrome mushroomed.

Unlike trends in other global(izing) cities, particularly as documented by Zukin in New York City (1982), inner Sydney warehouse occupations may have left few, or no official records but, as Horvath and Engles (1985, 146) did notice that '[t]he conversion of warehouses and light manufacturing structures into 'loft-style' residences . . . appeared in Sydney in at least the early 1980s'. However, this activity was largely overshadowed, in scholarly or governmental terms, by interest in other processes of gentrification, particularly of old inner-city *residential* areas. Where documentation exists of former commercial/industrial landscapes being transformed for residential purposes, the focus tended to remain on more classical gentrification analyses, which emphasized economic factors and associated issues of formally-acknowledged displacement. In the now gentrified dockside suburb of Pyrmont, for example, pockets of 'working class' residents were displaced as their homes, mostly old terraced houses, were torn down with the almost complete redevelopment of the area (Kendig 1979, Fitzgerald 1987, Fitzgerald and Keating, 1991, Frost 1992, Howe 1994 and Bonyhady 1995 on redevelopment in Balmain). The other parts of Pyrmont, the industrial and port areas, were also redeveloped but the commentaries have come more from 'boosters', those with a vested interest in selling such properties, rather than from the quarters of careful socio-cultural analyses.

Before Pyrmont was disembowelled, and reborn, individuals or small groups bought and converted some of the smaller warehouses and factories around the inner city. On occasion, former 'illegal' occupants-bought the buildings in which they had lived. As previously mentioned, a few buildings maintained their 'informal' warehouse residents until they were redeveloped, however, far from what Zukin (1982) identified in Manhattan, which acknowledged a phase of artist-driven preservation and improvement of formerly run-down city spaces in Manhattan's SoHo, there was no 'historic compromise' between culture and capital in 'the urban core', in Sydney. Unlike what followed in New York, Sydney's first-wave of warehouse occupations, of renters and squatters, was not legitimized (or commodified). The Sydney scene remained 'illegal' and underground, and was then abandoned and largely (and deliberately) forgotten. However, this part of a city's history did set in train an appreciation for industrial chic, as well as the preservation of buildings. Warehouse living provided a foundation for what was to follow.

The rise of urban(e) chic

One of the earliest official warehouse conversions occurred before the recession of the early 1990s. This was The Watertower in Redfern, which was completed in the early 1980s and has remained a landmark expression of formalized residential 'industrial chic'. This site now serves as a gateway to the multi-functional, postmodern Australian Technology Park,[11], and the apartments built within are unique for their size – they are spacious. The next phase of apartment developments such as Sydney Mansion, was advertised in early 1998 as 'Warehouse Style Loft and Level Living... Ultimate city living in heritage building', and began to fetch prices that were well above the cost of a terrace house in the same areas. (Sydney Mansion advertised prices *starting from* $215,000 for a one-bedroom apartment, $300,000 for two bedrooms, and $420,000 for three in early 1998). The premium prices that this 'new style' began to attract were a sure indicator that a there was a market, and it was affluent enough to make taste choices (Bourdieu 1984).

Then, all of a sudden, Manhattan references began to creep into promotional discourses, and the term 'loft' began to overshadow 'heritage'. There were new-build apartments springing up but this was still a transition phase, and a clever method was required to promote these new apartment buildings. One tactic used to provide a bridge between old and new buildings was to was to feign 'heritage' in the promotions of these apartments.

The Benedict, 'Chippendale's finest warehouse conversion' was advertised in mid-1997 and was considerably cheaper than other areas, such as Surry Hills, at the time. In advertisements in the *Inner Western Suburbs Courier* (10, 16 June 1997), Chippendale was described as 'the "new Paddington" area of Sydney'.[12] Marketed by the same company as The Benedict, and situated next door, a similar advertizing campaign was used to promote The Buckland. However, unlike The Benedict, The Buckland was a new-build development. From the way the advertisements promoted both properties together, it was easy to assume that both were warehouse conversions. The glossy brochure emphasized the 'heritage sympathy' of the development. 'Even the bricks took three months [to make]; they're a special colour... chosen to blend perfectly with the surrounding warehouse colourings of adjoining buildings'. The cultural capital of heritage was deemed necessary to imbue the new site.

Gradually, however, the rusticity of industrial heritage began to fade in the discourses of apartment promotion. The new banner of urbanity was proving to be a useful addition, and a new focus was borne. In 1998, Warehouse 82 in Chippendale was advertised as an 'exquisite warehouse conversion', with apartments available for purchasing 'off the plan', that is, prior to construction. The expensive promotional kit featured a collage of inducements for buyers seeking 'urbanity' (postcards, cards with details about fixtures, rolls of architectural plans, and a large, spiral bound promotional booklet with a metal cover). The words and pictures emphasized high ceilings, an urban skyline, the ('heritage' and 'urban') location, consumption at the nearby 'Lights and Action on Broadway', and a living environment that was 'urban cool...leafy, urbane, arty, academic [!]'. The banner on the building shell simply stated 'True warehouse conversion.' The Savoy On Shepherd was then advertised, just around the corner from Warehouse 82. The asking prices were very high for the area at that time (1998): studios (bedsits) started at $199,000, ranging to three-bedroom apartments priced at $595,000, which was on a par with the median price for a terrace house in Paddington, the benchmark of upmarket gentrification, at that time.[13] The Savoy was advertised as 'the lifestyle of the City...Fully refurbished historical building...22 Spacious Loft Apartments and 4 Elegant Studio Apartments'.

With an increase in developer interest, and a finite stock of old industrial buildings, the new phase of apartment development in inner Sydney could no longer simply invoke 'industrial heritage'. The 'New York style loft apartment' arrived in the late 1990s, when a renewed self-consciousness emerged as the global spotlight turned to the city of Sydney for the 2000 Olympic Games. Quietly stirring were some old, well-embedded imaginaries from the city's past: the dream of Gotham was being recalled.

Architectures of Escape 2: Sydney's SoHo Syndrome

> The Manhattanisation of Sydney proceeds apace...As the central city workforce becomes more affluent...linkages between global cities – Sydney and New York...– begin to rival linkages between nation states. (*Sydney Morning Herald*, 27 February 1999, S1)

Apart from such bombastic journalistic claims, visual and other references to Manhattan increased around the city of Sydney in the frenzied lead-up to the 2000 Olympic Games. On the bustling foreshore, where the CBD meets the harbour in the shadows of Sydney's famous Opera House and Harbour Bridge, a lone street performer provided Circular Quay with its own, modest, live human sculpture of the Statue of Liberty. In early 1998, a series of advertisements featured on three of the most prominent billboards in the city, placed to catch the attention of Sydney's main road users. One dominated the southern exit from the Sydney Harbour Bridge and captured the attention of southbound and westbound traffic; the second loomed over Taylor Square, a major arterial road junction for vehicles travelling east and south; the third looked down on Broadway (yes, Broadway), catching the attention of those travelling out of the CBD to the west or to the south. The advertisement for an airline featured an enormous image of the Statue of Liberty with the sails of the Sydney Opera House replacing the usual flames atop the statue's torch. The image cleverly melded two cities: New York and Sydney. Succumbing to the inevitable temporary life of billboards, the advertisements were eventually removed but a much larger version appeared in early 1999. It was placed at eye height for commuters who cross one of Sydney's more recent icons, the Anzac Bridge.

At first glance the advertisement appeared to be a marriage of the bigger older city (New York) and the smaller showpiece city (Sydney). A famous New York icon (Statue of Liberty) held a famous Sydney icon (Sydney Opera House). A momentary glance, such as flashing past in a car, gave the quick and simple message that the airline being advertised was bringing New York and Sydney together. A more sinister reading was that the 'big sister' city might be incinerating the little city. Regardless of the possible interpretations, the Sydney-based advertising company consciously employed an unmistakable New York referent. For those traversing Sydney, by road, the airline advertised was offering transport to New York City. Meanwhile, away from billboard advertising on main roads, real-estate promoters were promising to bring New York to Sydney, perhaps *turn* Sydney into a quasi-Manhattan (metaphorically igniting Sydney with that torch!).

Lofting the urban psyche: Sydney's SoHo Spectacular!

To appropriate is to take control over that which originated elsewhere for other semiotic/ideological purposes. (Collins 1995, 93)

New York lifestyles may be promised in real estate sales-pitches in gentrifying inner cities around the world, but the Sydney scenario became quite literal, as we shall see). Suddenly, new developments carried names such as The Manhattan, The Madison, The Lincoln and Tribeca. The Dakota development mimicked its (in)famous Manhattan counterpart rather than referencing any 'wild west' US origins (Figure 4.1), and the apartment block dubbed On Broadway was advertised utilizing its proximity to the bustling new shopping complex The Broadway

Figure 4.1   The Dakota

(promoted as 'The Greatest Show On Earth'). The Broadway occupies a complex of restored heritage buildings, which feature two huge clock turrets topped with enormous glass globes. This landmark site is located on Broadway, a main thoroughfare to the CBD of Sydney. Another apartment development, Off Broadway is just off to one side of Broadway (as the name suggests). The advertising campaigns of many of these new developments liberally used images of the Empire State Building and Statue of Liberty in promotional materials (Figure 4.2).

The property-market activity in inner Sydney, at that time, can be identified in the spiralling cost of property. House prices in the inner city increased by 19 per cent between 1993 and 1996 (compared to 9 per cent and 8 per cent in middle and outer city zones), and apartments increased by 24 per cent over the same period (compared to 3 per cent and 4 per cent increases for middle and outer city zones) (Daly 1998, 60). The NSW governmental push for urban consolidation, and the strengthening vision of Sydney as an international/global city, had

Figure 4.2   Imagining New York, New York

enmeshed with international trends of corporate-driven re-development of inner cities (Sassen 1991, Daly 1998). A range of factors had ensured that there was a (perceived) need to house more people in inner Sydney, and the 1996 Census recorded a reversal of the 80-year trend of decreasing inner city population. The residential population of inner Sydney had increased for the first time since 1911 (Daly 1998) with increases of approximately 16 per cent per annum between 1991 and 1996, and then by approximately 16.5 per cent during 1996 to 1997 (Vipond et al. 1998). The latter authors also noted that since 1988 the inner city had experienced the greatest rate of increase in construction of multi-unit dwellings. Changing employment structures included a decrease in 'blue-collar' work and an increase in 'white-collar' work in the CBD, and inner city. All these changes have contributed to providing a market for the promoters of 'New York style apartments', and these developments provided new spaces of consumption (of heritage and/or urbanity) in the postmodern residential city.

The following is a detailed account of the emergence and consequent installation of a distinctly New York/Manhattan narrative for Sydney's new residential landscape.

Promoting New York Style loft apartments

An obvious way to identify the New York theme in new developments/conversions was to note the tendency to adopt specific New York nomenclature in many advertising campaigns. Through the late 1990s in inner-city Sydney, Manhattan names and other identifiers of the building of a Manhattan 'loft apartment' fantasy were used in Sydney apartment promotions.

An editorial in the *Glebe & Inner Western Weekly* (3 September 1997), conjectured that the first New York theme apartment development was the Central Park warehouse conversion in Chalmers Street, Surry Hills, with '70 "loft-style" apartments' (a development with a confusingly similar name is the newer Central Park Apartments, in Chippendale, promoted during September 1998). One of the earliest 'loft conversions' was actually Lacey on Regent, in Redfern, which was promoted in February 1997. Lacey marked a notable change in real-estate promotional technologies. A huge banner was slung from the top of the existing building in Redfern, in which the development was to take place. The banner resembled one of the enormous advertisements on the sides of Manhattan buildings in Times Square or Houston Street, which separates SoHo from the more central Greenwich Village district.

Slung from the building, a larger-than-life photo of a woman, carrying with seductive composure the caption 'When Size Matters' (obviously a reference to the size of the lofts) loomed over Regent Street. To assure prospective buyers that the otherwise stigmatized Redfern area would not be problematic, a section of the advertising brochure was dedicated to a discussion of 'security', and displayed a photo of a security camera. The brochure also assured prospective buyers that Lacey on Regent was 'Sydney's hot spot of growing value . . . a hot investment' in an area that might be the next Paddington.

Throughout the inner city of Sydney, direct New York referencing began with a spate of loft-style developments with distinctly New York names (see Box 4.2). The first overt New York name was, in fact, The Manhattan. These 'loft apartments' were advertised in the *Sydney Morning Herald* in April and May 1997 (and at various other times since such as in June 1998). Included in the right-hand corner of the advertisement was a shadowy image of the Statue of Liberty. The Madison, referencing Madison Square Gardens in Manhattan, promised a 'Manhattan Lifestyle', in Chippendale, Sydney, in April 1997. Around the same time Dakota was advertised. An accompanying photo of the development echoed (somewhat loosely) the design of the Manhattan original (Figure 4.1). The advertisement stated: 'The new Manhattan style Dakota apartments offer the vibrance and excitement of the city . . . at your door. . . . video/intercom security and parking/highspeed lifts'. This Sydney version offered a formula for a dynamic Manhattan-style apartment, complete with high security (presumably, death by gunshot would therefore be highly unlikely).

Dakota was one of the many Meriton developments in the city. Meriton is a large Sydney-based property-development corporation and followed the Dakota development with the release of Paragon, The Paramount and The Palladium and used explicit New York referencing. 'Dalgety Square: New York Style! . . . a reconstructed heritage building' was advertised in July, 1998, with 'New York style living in Sydney's thriving heart'. In April/May in 1997, Meriton started prominent full-page advertising in the main real-estate promotional medium: the Saturday *Sydney Morning Herald* Real Estate section, now dubbed 'Domain'. Meriton had cornered the market in residential city tower developments in the Central Business District of Sydney, as well. The New York theme prevailed with stepped rooflines, use of two-tone colour schemes, and echoes of 1920s and 1930s Manhattan architectural styles. These developments include The Windsor in Kent Street, Millennium Tower on the corner of Sussex, Day and Bathurst Streets,

Box 4.2 Manhattan dreaming in Sydney

Building name	Advertising 'hook'	Year	New York City reference
Watertower	'property of the week'; 'sensational warehouse apartment'	1982	industrial heritage; 'SoHo'
Chelsea	apartments; imagery of NYC	1999	NYC neighbourhood
Printery	'loft apartments'	1997	loft; SoHo
Lacey	'loft conversion'; 'LeCorbusier style loft apartments'; 'avant garde interiors'	1997	SoHo, and architect
Manhattan	'loft apartments'; 'new two storey loft apartments' (Statue of Liberty illustration)	1997	Manhattan/NYC
Madison	city park/garden parallel in advertising copy	1997	NYC landmark
Dakota	'new Manhattan style . . . apartments'	1997	NYC landmark
Paragon	'elegant living, New York style'	1997	NYC landmark
Palladium	New York Style high-rise illustration	1997	NYC landmark
Paramount	'New York Style Loft Apartments'	1997	NYC hotel
Dalgety Square	'New York style living in Sydney's thriving heart'; 'historic heritage building'	1997	NYC landmark
Powerhouse	'classic city living elegance of New York'	1997	NYC reference
Lincoln	''New York' style apartments'	1997	NYC landmark
Union Square	'an absolutely authentic warehouse in the grand style!, New York style'	1997	NYC park
TriBeCa	'epitomises New York warehouse style'; 'Slice of New York with strong Sydney flavour', ''loft-style' apartments; 'birdcage lift'	1997	NYC neighbourhood
Benedict	'Chippendale's finest warehouse conversion'; 'loft apartments'	1997	loft; SoHo
Buckland	'heritage' (advertised with The Benedict)	1997	loft; SoHo
Mark	'warehouse apartments' (lofts); Get in on the Act; new star on Broadway'	1997	Broadway, NYC

Name	Description	Year	Location
On Broadway	'the investment all Sydney is applauding'	1997	Broadway, NYC
Off Broadway	'the all singing, all dancing, bright lights, great life apartments'	1997	Broadway, NYC
High Holborn	'warehouse apartments'(lofts)	1997/8	loft; SoHo
Harrington Grande	'New York lifestyle' (Sydney heritage)	1998	NYC
Cyclops	'stunning warehouse apartments' (heritage)	1998	loft; SoHo
Piano Factory	Heritage	1998	loft; SoHo
Sydney Mansion	'warehouse style…heritage building	1998	loft; SoHo
Savoy	'spacious loft apartments'	1998	loft; SoHo
Metro	'the Warehouse Erskineville via New York'	1998	NYC
SoHo Apartments	'the centre of style'	1998	loft; SoHo
Valentino's Apartments	'exclusive loft apartments' (Liza Minnelli in 'Cabaret' look-alike on brochure)	1998	loft; SoHo
Warehouse 82	'exquisite warehouse conversion' (lofts in heritage building)	1998	loft; SoHo
York	' 'New York, New York' loft apartments'	1998	NYC; loft; SoHo
Central Park Apartments	'style, design…sleek…warehouse…the New Era'	1998	NYC landmark
Hordern Towers	'exciting urban dream'; 'out of this world'	1998	Gotham image
Elizabeth House	'Art Deco…the latest in New York style city living'	1998	NYC deco
The Tower	'Fifth Avenue' (huge b/w Gotham ad.)	1998; 2000	Gotham/NYC landmark
Carlisle	'something old…converted warehouse…lofts'	1999	NYC landmark
Hudson	4 stage development: The Met; Astoria; Lincoln; Carnegie	1999	River, NYC landmark
Stijlvol Zone	'Rem Koolhaas OMA designed' in 'old Mark Foys warehouse'; 'all within a unique heritage façade'	1999	architect & NYC
Waldorf	luxury apartments	2000	NYC landmark
New Yorker	'make your debut off Broadway'	2000	NYC, Broadway

Campbell Tower, Castlereagh Tower and The Regis Tower on Castlereagh and Campbell Streets. The New York theme was recognizable in these developments even without Manhattan-referenced names.

A spate of promotions continued throughout 1997 with 'Elegant living, New York style' in the 'The Powerhouse Apartments'; The Lincoln offered 'New York style apartments with soaring ceilings and polished timber floors' (Lincoln, of course, referenced the famous Lincoln Centre, a major Manhattan landmark). Union Square was advertised as 'an absolutely authentic warehouse conversion in grand style! . . . The latest expression of loft living New York style in the vibrancy of Sydney', in June 1997.

Taking the SoHo Syndrome to a new level, TriBeCa burst on the scene in mid-1997. Located in Chippendale TriBeCa promised 'Warehouse living that exceeds your expectations' and, 'TriBeCa epitomises New York warehouse style . . . a birdcage steel lift . . . unrestricted skyline'. The promoters were well aware of the New York connotations of naming the development after a Manhattan neighbourhood that was also in the throes of loft development: 'TriBeCa, which takes its name from a fashionable area in lower Manhattan, New York' (*Glebe & Inner Western Weekly*, 3 September 1997). The sales brochure also noted that 'you can keep an eye on your neighbourhood through large windows . . .'

Just to the north of the Redfern area, in Ultimo, a warehouse apartment development called The Mark was advertised in November 1997, using the heading 'Broadway'. The later advertisement stated 'Get in on the Act: New Star on Broadway', utilizing another take on the New York theme.

Suddenly the theatre district of Manhattan became the theme for anything near Sydney's own Broadway, a main thoroughfare to the CBD. In September 1997, On Broadway had its 'Grand Opening: the investment all Sydney is applauding'. 'A choice of superb . . . lofts at a price that's a steal on any other show in town . . . at the heart of the city's restaurant and entertainment belt' could not be more explicit. Seizing the opportunity, the promoters of Off Broadway, in Chippendale, produced a glossy brochure with the words 'Off Broadway' surrounded by a circle of yellow stars, with a photo of a model of the development lit by spotlights, which resembled a stage. The caption below read: 'The all singing, all dancing, bright lights, great life apartments'. On the flip side were three stills from black-and-white Hollywood movies from the 1940s/1950s.

Following the Broadway Spectacular theme, Valentino's: Apartments on Crown (*Sydney Morning Herald* 30 May 1998) featured a Liza Minnelli (in the film *Cabaret*) look-alike in the advertisement, and photos of slick, minimalist loft apartments. 'Exclusive loft apartments...dazzling lights...a tantalizing way of life, as distinctive as you are...or want to be', provided a less explicit theatrical reference. This development was built in the heart of Sydney's pink triangle precinct. A full-page advertisement, carrying the text:

> Start spreading the news...I'm Leaving Today...I'll make a brand new start...If I can make it there, I'll make it anywhere!...I want to be a part of it...It's up to you, new York, new York,

followed the Broadway spectacular theme for the 'New Release...[of] A new YORK Apartment.' (*Inner Western Courier*, 7 September, 1998).

Reaching a new level of competitive absurdity, the Metro in Erskineville, just south of Redfern, offered *everything* in March 1998. The glossy brochure advertised:

> [a] warehouse conversion apartment [building, with the] best of old world architecture and space with the ultimate in today's conveniences...cosmopolitan...video intercom...security parking.

This complete lifestyle package was advertised under the banner 'Metro – the Warehouse Erskineville via New York', which blended location(s) with industrial chic, all under the watchful eye of surveillance technologies. The Manhattan/New York nomenclature continued with SoHo Apartments (*Sydney Morning Herald*, 11 April 1998), 'The centre of style', and the Central Park Apartments (*Inner Western Suburbs Courier*, 17 September, 1998).

Arguably the most spectacular and exclusive Manhattan statement and development was advertised in the *Sydney Morning Herald* in early 1998. The advertisement was for The Tower, built in the commercial heart of the CBD. 'The city's most exclusive residential offering', featured a long narrow, full-length broadsheet advertisement. The image was a distinctly Gotham City caricature of Manhattan. Later advertisements (*Sydney Morning Herald*, January 1999) evoked 'the vibrancy and pulse of a Manhattan lifestyle', and the same Gotham City image then appeared in atmospheric blue ink, with the words 'the Tower Sydney...5th Avenue Residences' (*Sydney Morning Herald*, February 1999). Residential Manhattanism had become part of the character of Sydney's CBD.

Within the Manhattan narrative, the recurring themes were metropolitanism, cosmopolitanism, spaciousness (soaring ceilings), open (plan) style, access/location to the city and services, city views, and occasional

European 'style' references (stainless steel and clean lines are associated with Europe in some promotions, and SoHo lofts/industrial pasts in others). Occasionally industrial heritage was mentioned, but the Manhattan narratives dominate. New York-ness had become completely embedded in this Sydney-specific imaginary of city living.

Consuming (loft) apartments

By the end of the 1990s, 'loft living' had become a well-entrenched housing option for Sydneysiders, particularly for those with the financial power to put economic necessity at a distance.[14] One new loft resident put it this way:

> Basically the apartment made my mind up... I was searching in the inner city, and when I saw this, I just chose... this apartment. [It was] not price related, it was a little more over-priced than other places but... this building is amazing! (Interview, November 2001)

Once the cultural capital of industrial chic was completely embedded, the word 'warehouse' was easily replaced with 'loft' in promotional discourses. The lifestyles associated with lofts grabbed the media's attention. 'Never mind the unit – it's a lifestyle, darling' (*Sydney Morning Herald*, 20 April 2000)[15] ensured the burial of the old notion of the tired old 'home unit' or 'flat'. The new housing form – the lifestyle apartment – was fast becoming a 'must have' for some, as one loft resident remarked: '[This is] a brand new building. [When] this came up and we were like 'get in' [to the latest trend]' (Research Interview, November 2001). And a CBD resident summed it up this way:

> We liked the modern idea of the apartment. We thought the view was great – the fact that it could never get built out – because this is all heritage... all sandstone building.... It's north facing... you have to get that sun. It had car parking. It was designed by a reasonable architect so that was all right. And the builder was quite a good builder... But when we bought this apartment it was fairly instant... we fell in love instantaneously, because it had the north and the east... it had good ventilation, high ceilings. (Interview, April 2005)

By the late 1990s, Sydney's 'New York style apartment' had evolved to a point where references to warehouse heritage were no longer required. However, as detailed earlier, a mandatory accessory for this new urbane style of living – highlighted throughout the marketing rhetoric for apartments – was the inclusion of fortressing technologies. One resident, reflecting on the issue of security, remarked:

> It's a great area [but] it has its problems with security, there's no question about that, but once you know what to be careful [of]... I mean there are certain things you can do to prevent danger... certain things you can control. (Interview, November 2001)

A CBD resident put it this way:

> We can lock it up and just walk out the door and it just feels secure. You know, living in a house... if you heard a noise... you might get a bit frightened that it was somebody lurking around the garden or whatever. Here, it's really private, and, it's fairly high – no, I would call this really high security... we've got security doors, concierge, you know, so we feel really safe here. (Interview, April 2005)

For these loft/apartment residents, the capacity to 'control' elements of danger, through care and prevention, combined well with location. Along with the issue of security, the themes of convenient access to the city and public transport were also well rehearsed during research interviews).[16] Another resident noted an extra dimension to the advantages of location: 'I like [it here] because it sits on the crossroads of the major roads...I spend most weekends away... and you can hit all the major expressways [easily/quickly] (Interview, November 2001).

When not escaping the city, these city-dwellers know that their vehicles are securely locked away, below their apartments, and are (usually) accessible via elevator or stairwell linked directly to the basement parking lot. Urbane lifestyles can proceed unhindered by events outside. As one 'loft' resident remarked: 'Some people have said that it's quite a bad area... for me, no problems at all, even if it's two minutes from "The Bronx" [Redfern]' (Interview, November 2001). Within Sydney's SoHo Syndrome, the physical and metaphorical removal from the realities of the street provides refuge from the presence of danger.[17] Any engagement with the local area is by choice, not necessity. The attributes of fortressed urbanity are part of this version of cosmopolitanism.

In areas stigmatized by the close presence of The Block, promotion of property fortressing has assisted in the establishment of beachhead developments. These highly secured apartment blocks have wedged open spaces for the colonization of an otherwise 'no-go zone'. Although danger is now largely avoidable for these apartment dwellers, it can serve another purpose, as observed by a resident of an early near-Block conversion:

> When [we] first moved in here...we walked [at night] down through...['Redfern']. I knew there was a degree of risk attached...I've lived in a number of cities in the world that would only be considered highly dangerous...I was pretty alert and astute to what was going on. But I did it as an exposure to myself, and particularly...because [my partner] really wanted to. (Interview, February 1998).

In the city's theatre of danger, fear can become entertainment and, perhaps, an education. Although not a pursuit that appeals to everybody, one real-estate agent did mention that the presence of The Block, 'the ghetto' added a little 'culture and colour' to the area, it provided a focal point or a 'special interest' feature (Interview, July 1998). For some, an edge of danger is part and parcel of city living (Smith 1996). It also adds a little extra to the smorgasbord of cosmopolitan offerings.

Away from the potentially dangerous and sometimes thrilling local area, new residents can relax, and disengage in their upmarket/upscale 'New York style apartment'. According to Simmel (1995 [1903], 35):

> The essence of the blasé attitude is an indifference toward the distinctions between things.... What appears...as dissociation is in reality...one of the elementary forms of socialization.

For those with the economic capacity to choose to live in a city apartment, urban living has come with the offer of new cultural formations. The kinds of urbanisms on offer can be imagined to exist in any urbane city, around the world, where the hum and throb of the street exists outside, below the sanctuary of the apartment. These new urbanisms have captured the attention of Sydney's property consumers, be they new residents or savvy investors. One resident summarized the attraction:

> [The apartment had] space and...good light, and the area appealed to us...its urban, recycled warehouses...that sort of thing...easy walking distance or easy transport options....[The apartment] had what we wanted. (Interview, November 2001)

The commercially manufactured urbanisms associated with apartment living in Sydney come with a range of socialized capacities. These include the ability to engage in safe, urbane and convenient lifestyles that can deny, be indifferent (or blasé) to aspects of the local, its histories and its pathologies. When desired the local can be engaged with for entertainment or consumption purposes. These sorts of urbanisms are now embedded; they are a part of Sydney's SoHo Syndrome. The promotions of these ways of living, in the increasingly exclusive and exclusionary cosmopolitan metropolis, continue to this day.

Conclusions

> Manipulated ... to represent the current social hierarchy as natural and permanent ... we may speak of urban form as an ideological project. (Kearns and Philo 1993, 13)

This chapter has detailed a recent transformation of inner Sydney into landscapes of 'New York style apartments'. By invoking imaginaries of Manhattan, the specificities of the local and its context have been easily forgotten, or ignored. I have used the example of 'forgetting' a particular history, in the accrual of cultural capital, to demonstrate how local experiences are linked to broader political economies and identity formations. New apartment/loft-dwellers have no reason to acknowledge that which has gone before or exists currently in the local, in the consumption of the Manhattan-like lifestyle (and sales) package. Once disengaged from the pathologies of the externally present city the 'protective blanket' of denial and indifference that has been embedded into the lifestyle package of apartment/loft living, can been fully operationalized. New apartment-dwellers can deny, be indifferent to, and/or consume as they wish. The comfort zone, away from the world outside, has a range of offerings for the new cultural consumers of the city.

The presence of The Block may have complicated the fantasy of a clean and 'white' Manhattan produced by the marketeers of apartment developments. However, everyday practices of denial or indifference to social inequalities and the pathologies they create have also been socialized through discourse, and landscape production. Alongside urban lifestyles that have included sophisticated consumption practices (such as café cultures and multicultural culinary experiences), the postmodern metropolis cannot completely hide its underbelly. Although the lives of others have been, on occasion, commodified as part of consumption choices, the economic and social reality is that as the experiences of the new apartment/loft-wellers and the residents on The Block polarize, they will continue to abrade each other.

The advent of apartment/loft landscapes in Sydney in the late 1990s was the embodiment of global *and* local, economic, cultural and political processes that have provided property developers with large-scale opportunities. At the same time, they created new and exciting but increasingly exclusionary forms of urbanisms, or ways of living in the city. Sydney's own histories, and the 'cities of differences' (Fincher and Jacobs 1998) – the earlier warehouse occupations, the Indigenous battles for recognition and territory in the city, the city's multiple ethnicities, sexualities and physical/mental abilities – have found no expression except as consumable offerings in the cosmopolitan metropolis. The production of apartment/loft living in Sydney may have reoriented the style and function of the city in spectacular ways. It was, however, the specificities of inner Sydney as a locale, as well as the complexity of issues of identity that were promoted and not promoted, bypassed or consumed as part of a city's cosmopolitan urbanism that remains ever-present within Sydney's SoHo Syndrome. The promotions of apartment living have offered a generic, or imagined form of 'white' Manhattanism than can be replicated in cities around the world. In the example of inner Sydney, identity formations have been tied to the fantasy of globalized white identity formations (Bonnett 2000). In the manifestation of Sydney's SoHo Syndrome, whiteness has been relationally constituted and variously expressed through the built form.

Notes

1   I am using the most generic definition here, which means that these former industrial sites were not necessarily contaminated, although many probably were.

2   I have drawn on the idea of 'dormitory suburbs', and used the term 'dormitorization' to indicate the shift from commercial/industrial uses of urban spaces, and built environments, to sleeping quarters.

3   The 1996 Census found that 'a market increase in medium to high density housing [occurred in] . . . the City of Sydney and nearby inner suburbs' (ABS, Sydney, *A Social Atlas*, 2030.1, 1996, 14). The *Sydney Morning Herald* reported that 'The City of Sydney Local Government Area . . . the fastest growing municipality. The population base grew by 25.2 per cent in the year to June 1998' (*SMH* 28 April 1999).

4   The term 'loft' has been used very loosely in Australia. On occasion, mezzanine levels exist in old or new buildings. At other times, 'loft' implies that the apartment is housed within an old 'converted' building. At other times, 'lofts' bear no resemblance to the housing form of the same name found elsewhere.

5   From *Crocodile Dundee* to recent films like *Getting Square*.
6   Zoning regulations had changed to enable warehouses and old office blocks to undergo residential conversion, and many industrial/business zonings changed to Zone 2b, 'medium density residential', and Zone 10, 'mixed use'.
7   Margaret Rapis, a former media and communications student at the University of NSW, produced a CD-Rom memento of Alpha House. According to Rapis, Alpha House was 'once a haven for the creative, [it] is one of the wounded prey of the economically rational status quo'. (http://mdcm. students.arts.unsw.edu.au/click/alpha_house/preview.html). The new development has retained the Alpha House name.
8   From discussions with warehouse survivors about (our) personal experiences of 'the Slaughterhouse' in the mid- to late 1980s and again in the early 1990s, which included reminiscences about the characters and bands that practised and performed there. For a list of Australian bands from that era see http://www.innercitysound.com.au
9   I sighted the plans to convert Beta House into apartments at (the former) South Sydney City Council's Planning and Building Department.
10 From the late 1980s to 1995, the Sydney 'rave' scene also used these warehouses.
11 The Australian Technology Park (ATP) is a joint venture of three large Sydney universities. It acts as 'a hub to link their world-class research facilities to industry and the wider community' (ATP homepage, http:// www.atp.com.au/whatis.html). It is located at the old Eveleigh railway yards adjacent to Redfern Railway Station.
12 Paddington is *the* benchmark suburb of inner Sydney gentrification.
13 Residex figures for March 1998 indicted that the median house price for Darlington and Chippendale, was $290,000.
14 In Australia, the number of people living in high-rise units rose from approximately 129,000 in 1981 to around 334,000 in 2001, with more than half (68 per cent) of all high-rise residents living in New South Wales (Australian Bureau of Statistics, *Australia Now*, 2: *Australian Social Trends, Highrise Living Report*, 15 June 2004). Between 1994 and 1999, 12,970 Development Applications (DAs) for 'residential projects' (apartment buildings) were approved for construction in the City of Sydney Local Government Area, with 7,330 completions. Between 1999 and 2004, 6,329 DAs were approved, and there were 7,257 completions (City Residential Monitor, City of Sydney, 2000, 17, p. 3; 2004, 35, p. 3).
15 Other examples include: 'Fabulous homebodies: marketing the urban dream', *Sydney Morning Herald* (19 June 1999); 'Step up to life in the Style Zone', *Sydney Morning Herald* (17 July 1999); 'Movable modernism: a Sydney design duo creates a buzz from New York', *Sydney Morning Herald* (26 April 2000).

16 See Lees and Shaw 2002; W.S Shaw and A.L. Semple, Manhattanising Sydney. Preliminary unpublished report, University of New South Wales, 2005.

17 For most of the residents of luxury apartments interviewed in 2001, local issues were subordinate to the broader economic/geographical issues, such as cost of property versus convenience/access to the CBD.

5

Cities of Whiteness

Geographies of Urban Whiteness

It does not take incredible powers of observation to notice obvious forms of racialization. For me, the act of travel, arriving at Heathrow Airport in London, or Los Angeles International, is an instant reminder of how obvious racialization can be (in this instance through divisions of labour). I am always surprised by such immediacy, that airport staff *appear* to be assigned jobs by designations of skin colour in ways that happen less visually – from my observations at least – at Australian airports, or in this country overall, which is why I continue to be taken aback by this. With such overt visual categorizations it is very easy to think of racialization as obvious, and linked to biologically derived distinctions such as skin 'colour', with 'white' often presiding at the top of such hierarchies. By such logic, this kind of visual evidence promotes the understanding that experiences of racialization occur less frequently in places where the distinctions are less visually obvious. Because this is highly unlikely to be the case, the Australian experience (though not necessarily unique in this regard) has provided a context for unearthing some of the other, less overt ways that racialization processes of whiteness operate. In the preceding chapters of *Cities of Whiteness*, I focused on examples of whiteness that have occurred within the less traversed terrain of urban transformation, where very specific cultures of privilege have emerged in the shadows of a more obvious and attention-grabbing landscape of 'racial' segregation, in inner Sydney. Although existing conceptualizations of urban change have engaged with the issue of racialized segregation, my task for this book has been to problematize and complicate the concept of 'whiteness' within the analytical field of emergent cosmopolitan urbanism.

I have unearthed a series of urban cultural processes that have worked within renewed cityscapes, where the exclusionary and malevolent aspects of the emergent urbanisms have remained largely hidden behind the unexpected, and visually overt segregation that exists in inner Sydney. The presence of the Aboriginal settlement known as The Block, where ('unworthy', and demonized) 'black' confronts and frightens (entitled, and normalized) 'white', has proved to be a handy smokescreen for processes that privilege. In this context, I have identified such processes as 'whiteness'.

A central tenet of this book has been to raise concern about the performance and consolidation of whiteness, as these have occurred through less obvious, but nonetheless powerful processes of material and non-material urban transformation. Although based in a corner of Sydney, where segregation is overt, I have built an analysis of the logics of settlement in the transforming (post-/neo-)colonial urban setting of inner Sydney, more generally. I focused my analyses on the re-establishment of particular kinds of neo-colonial identity formations that have manifested through the production of urban landscapes, and the rise of associated consumption practices in the new residential city. Increasingly, 'critical race studies' in Geography have sought to problematize privilege in order to make sense of processes that advantage particular groups. By focusing on examples of processes that privilege, this book has contributed to the now quite substantial scholarly move beyond the conventional study of disadvantaged groups, which has led to the study of whiteness. The analyses presented here have added a conceptual layer to this kind of study. I have sought to unpack the foundations for place-based understandings of 'white(ness)'. This book has detailed the ways that whiteness, as a set of privilege-making processes, is historically anchored yet remains flexible, slippery and multi-faceted in its operations. In *Cities of Whiteness*, I have engaged with and contributed to existing debates on the racialized politics of heritage, and urban regeneration, to mount a case for the interrogation of post-colonial whiteness.

The aim of this final chapter is to draw together the conceptual threads of *Cities of Whiteness*, and to suggest ways forward in the overall project of unravelling geographies of whiteness. To operationalize this aim, I draw together the book's core themes of *Cities* (detailed in the sections titled 'Urban Transformation and Gentrification' and 'Racialized Urbanisms'), and of *whiteness* (discussed throughout, but specifically in the section 'Particularities and Fantasies of Whiteness'). In previous chapters, I provided examples of multiply expressed amalgams of empowerments to demonstrate that whiteness is a highly complex set of processes. I have also sought to further an existing geographical project that challenges the assumed ethnicity of whiteness (which often ties 'white' to 'Anglo'). This understanding persists in scholarship and in common thinking in Australia and elsewhere.

The assumption of white ethnicity not only detracts from identifying more nuanced processes of whiteness, it can also serve to enhance the power of whiteness, as I have sought to demonstrate in the preceding pages. I have detailed how, in the case of inner Sydney, discourses that continually reinforce the sense of racialized segregation are part of (processural) whiteness. One of the outcomes of the relentless narrative, of the discourse of The Block's decline, is the reinforcement of unevenly bifurcated urban space and the reinforcement of similarly uneven understandings about Aboriginality and non-Aboriginality. This, in turn, has perpetuated and reinforced the long-held understanding about entitlement to urban space that has built on a history of actively excluding Indigenous peoples, from the moment of invasion and ensuing colonization. It also perpetuates the kinds of understandings about 'race' that began at that time. The neo-colonial circumstances of poverty and dispossession that haunt many Aboriginal Australians to this day remain hidden beneath the ongoing logic of Aboriginal inferiority, unpredictability and wildness. With the transformation of Sydney's inner city over the past few decades, and the reclaiming of what was a once undesirable urban space (except for those with few other accommodation options), new ways of living and consuming the city have emerged. The rise of exclusionary heritage impulses and protectionism, and the rise of cultures that avoid the here-and-now, are examples of activities that, although separately operationalized, work towards similar ends. Whiteness is not, however, one thing or outcome but its categorization as such has proved useful in creating an image of 'victim' (of crimes to person and property by Aboriginal perpetrators). At the same time, the critique presented in this book may appear to allude to an identifiable category of whiteness – as 'Anglo' gentrifiers, new apartment dwellers, or the new 'middle classes'. In reality, such groups are far from static. People defy being categorized in these ways. They are multiply identified and their identities are variously shaped, and reshaped, over time. Moreover, as I have shown, processural whiteness *can* include non-'Anglo' people. Whiteness also slips easily across class designations. Harking back to the teachings of Toni Morrison (discussed in Chapter I), whiteness can operate in circumstances that are completely external to 'Anglo' worlds, but often does so within the context of a wider dominant culture that identifies as 'white' (even if only when pushed to do so in an otherwise normalized realm of whiteness). What is more constant than ethnicity is the driving sense of entitlement, or aspiration for such entitlements enabled by whiteness. The stories told in the preceding chapters are about formations of racialized empowerment that are specific to one city. Although I hope that I have demonstrated the particularities of processural whiteness, it is also clear that the *kinds* of localized expressions of entitlement that create social polarizations in cities are not bound to inner Sydney.

Cities of Whiteness has detailed some of the specific ways that whiteness has operated during a dynamic phase of urban revitalization in inner Sydney. Methodologically, the details gathered and documented within these pages have built upon demographic trends – about changing populations with the movement back to inner Sydney through gentrification, and the increase in housing availability through apartment/condominium development (and a consequent escalation in property values)[1]. Upon this foundation, I have mapped out various discursive environments, produced by mostly non-Indigenous residents, the print and electronic media, police, real-estate marketeers, and others about The Block. Such details were found in a vast array of media reports, through in-depth research interviews, and within various government documents and historical records.

I also chased, captured and documented unpredicted events and the responses to such events by urban dwellers (such as resident activisms), the mass media and government agencies. I followed announcements about changes in government policies and reactions to these changes. These moments provided important empirical detail for this book. What I found was that the battles waged by near-Block residents were overwhelmingly driven by desires to preserve very specific entitlements. Some of these have manifested through the valorization, and protection of 'heritage', and have reinforced entitlements to place and space (physically and symbolically). As detailed in Chapter 3's 'The Good Old Days', the efficacy of resident activism around The Block is undeniable. The Health Department of NSW knows this all too well as it continues to confront highly organized campaigns against any proposal for a drug-related health facility (Sydney, like many other cities around the world has serious drug-use problems), in the gentrifying city. A glaring reality is that there is a desperate need for workable strategies to alleviate the tensions in this part of inner Sydney (where The Block continually rubs against the predominantly non-Aboriginal space that surrounds it). It is, however, clear that although the provision of such a facility[2] has the potential to alleviate some aspects of the city's drug problems (which are currently blamed on the residents of The Block), this initiative's success would simply detract from the current imaginary for Sydney's inner city. The power of resident turf protection to mobilize, and the over-riding (but somewhat covert) agenda of simply being rid of the Indigenous presence, has been successful in its attempts to stymie efforts to assist community cohabitation. At the same time, certain strategies that have *appeared* to promote cohabitation have done so at the expense of the Aboriginal community. One example was the resolution of a contest over the future of the Wilson Bros. Factory site and (the now defunct) South Sydney City Council's use of 'community consultation' in its planning process. Although done in the spirit of 'grassroots democracy', these kinds of democratic processes are not necessarily equitable since definitions of 'community' can be slippery, and potentially exclusionary (Sennett 1976). The fate of the Wilson Bros. Factory site demonstrated just how flawed 'democracy' can be, as the process slipped into yet another example of whiteness-in-action. In this instance, the definition of community defaulted to the 'ordinary' (non-Block) population of inner Sydney. As popular discourses continue to typecast Aboriginal people as unable to exercise self-determination in the city, The Block continues to be cast as the 'failed human experiment', which can therefore be disregarded. Local processes that seemed to be operating in the name of 'community' in the case of the Wilson Bros. Factory site included only those with the right credentials for membership.

Other eventful moments captured during the research process for writing this book included The Block's listing on the Australian Heritage Commission's National Register, and the advent of the Redfern–Waterloo Authority. On 25 October 2000, the Australian Heritage Commission announced The Block's addition to the National Heritage Register of Australia, as a site of Indigenous significance. According to the Commission, the site was 'a national symbol of an Aboriginal community's ability to maintain their identity in a city setting'.[3] Although this listing did not protect the Victorian houses, or other old buildings on The Block (the *site* is listed, not the buildings), it did mean that any subsequent development applications would have to be scrutinized by the Australian Heritage Commission's Indigenous Heritage Section. The Australian Heritage Commission[4] had acknowledged the centrality of The Block to the recent history of Aboriginal Australia, of land rights and urban survival. Although this heritage listing did not guarantee protection of the existing housing on The Block, it did mean that the site would remain as Aboriginal land.

However, in October 2004, the advent of the Redfern–Waterloo Authority unsettled this certainty. In the aftermath of the 2004 'riots' in Redfern,[5] the Redfern–Waterloo Authority was founded on the passing of a bill and subsequent Redfern–Waterloo Authority Act, 2004. The Authority unveiled its 'masterplan' for Sydney's inner city, which included the full redevelopment of public housing estates, and The Block. According to Sydney's daily broadsheet, *Sydney Morning Herald* (29 November 2004, 1, 52, 171):

> **Revealed: how Redfern will be reborn** (headline)
> The State Government has a $5 billion plan to redevelop Redfern and the surrounding suburbs that involves seizing control of Aboriginal housing on [T]he Block and letting private developers take over two thirds of the area's public housing estates … the redevelopment of the notorious Block would increase certain property values by 30 per cent.

Certainly, many local homeowners would be pleased to see the area finally redeveloped in this way. For the Aboriginal residents on The Block, and their supporters, the Authority's power to acquire land compulsorily is of grave concern. Somewhat ironically, even within the context of the current climate of heritage protectionism that has built on a history of taking a stand against large-scale proposals, local understandings of heritage simply do not include The Block regardless of its official listing. Although there is no commemorative plaque that acknowledges the listing on the site, local conceptions of heritage (as detailed in Chapter 3, 'The Good Old Days') suggest that this would be of little consequence. Although a step towards ensuring the security of The Block as an Aboriginal place in the city of Sydney, the impact that this heritage listing has had on the myriad ways that whiteness operates awaits further examination. Because Indigenous heritage is generally associated with pre-colonial authenticity, recognition of the heritage of this contemporary, part-assimilated, variously marginalized and dispossessed Indigenous settlement is difficult to envisage. The contrast to the surrounding built environment of Victoriana and converted warehouse 'heritage' does not bode well for The Block's placement on any local heritage-protection agendas.

Meanwhile, the evidence presented in this book suggests that along with the pathologies of the city, the heritage of The Block could fulfil a different purpose. It could become simply another theme in the field of consumption for new urban dwellers, with renewed calls for the site to become 'a cultural centre'. The extent to which it will become an

Indigenous theatre in the consumption landscape of the cosmopolitan metropolis, or subject to a new round of antagonisms over heritage, could be a somewhat moot point as the latest politics of place reveal the conflict in planning for redevelopment. One version proposed by the Aboriginal Housing Company (that administers and acts as landlord on The Block) is the Pemulwuy project.[6] The other is the aforementioned NSW state (Labor) government's Redfern-Waterloo Authority plan. According to a recent newspaper commentary:

> As confidential [NSW Government] cabinet papers advised back in 2004, 'if [T]he Block is not redeveloped (to reduce Aboriginal housing), the commercial benefits flowing … would face a substantial reduction … probably in the order of 25–30 per cent'. In other words it's not so much a racist government as a monetarist government, presuming on, and pandering to, a racist market. (Elizabeth Farrely, *Sydney Morning Herald*, 6 September 2006)

It seems that to attract further investment from a particular 'market' requires a reduction in Aboriginal housing on The Block. This proposal by the Redfern–Waterloo Authority to attract the right kind of investors has met with some resistance. A 'candle light vigil' (10 August 2006), brought commentary and support from a range of high-profile individuals, including Aboriginal celebrities and boxing champions Lionel Rose and Anthony Mundine. Another dignitary at the event was Tom Uren, Minister for Urban and Regional Development during the short reign of the Whitlam Federal Government (which provided the funds for the purchase of land for The Block).[7] Tom Uren is an avid protector of Sydney's public space and, interestingly, 'heritage'.

Ultimately, however, the urban-settlement fantasies in this part of inner Sydney rest generally on hitherto largely unacknowledged processes of whiteness. As detailed in Chapter 2, '(Post-)Colonial Sydney', the history of blight in the city of Sydney provided a needed space for Indigenous people to settle. It was also the setting for ignition of some very specific forms of racialization. The lack of recognition of Aboriginal people as citizens resulted in their exclusion from the forces of emancipatory politics that were active at that time. Quite conversely, other local residents, the local council and others viewed the formalization of Aboriginal settlement as an invasion, like a noxious facility, or a 'human zoo', rather than a group of people who – though 'different' – also needed homes and jobs in the city. Yet, regardless of the opposition to its formation, the settlement known as The Block was formalized and remains, against all the odds. Indeed, it provided a venue for the emergence of a very conscious 'black politics', in the 1970s. Less consciously, and in the present, it remains a thorn in the side of the Australian fantasy of an equitable and proudly multicultural society that gives everyone 'a fair go'.

The rise of resident activism that now exists in this part of the city of Sydney has its roots in the earlier working-class resistances to territory losses. Anchored in the battle for territory against the incursion of the nation's oldest university during its expansion, the earlier politics of class warfare then slipped into a reactionary politics of enclave consciousness with the threat of 'invasion' by the (un)known Aboriginal 'other'. Although Aboriginal people had resided in this part of the city at least since the 1930s, it was the provision of an official space of difference, the gift of land, that attracted resentment. The myth of 'Aboriginal privilege' (Mickler 1998) remains, and is a widespread phenomenon in Australia. In inner Sydney, the result is an ongoing, and ever evolving form of turf protectionism that not only excludes The Block but also works against its entitlement to space, and its continued existence. Much of the resident activism has mobilized to protect the entitlements that were gained with colonization. This heritage, of entitlement to land (now urban space) in a place originally deemed *terra nullius,* continues for those who identify with, or ascribe to the reinforcement of a dominating historical reading. Regardless of the ever-increasing realities of multiculturalism (in its most general sense) and the recognitions of Indigenous and 'migrant' contributions to Australia's evolution, a new phase of reinscribing a form of 'white' Australia has emerged and found expression through the built residential form, and new ways of urban living.

Within the new residential city, The Block remains an unresolved paradox that contradicts a belief in the value of tradition, where Indigenous Australians belong in the outback. By this logic, they simply cannot cope in the city. For the purpose of capital, the ongoing presence of The Block has hindered the expected gentrification trajectory; The Block's stigmatized status has created investment wariness. I have traced the ways that discourses of decline, perpetuated by the mass media and others, have contributed to the maintenance of a value system about The Block, as anachronistic, paradoxical and inappropriate in a globalizing city that reaches to its English past, and a Manhattan/North American future, for inspiration, and escape.

Feeding comfortably into the vision of the progressive globalizing city, came the boom in apartment developments that can encircle The Block, and promise escape from its unpleasantness, and dangers. The onset of apartment development in the old industrial areas of inner Sydney has created specific kinds of urbanisms, forged through the promotion of news way of consuming the city. The image of cosmopolitan urban living has easily elided the local. Just like 'the past' (with heritage dreaming), 'the local' has also become part of the consumption *smorgasbord*, to be dipped into, or not, as dictated by whim. This has added another cultural dimension in a reconfigured cosmopolitan metropolis where issues of equitable access to housing have been relegated to the realms of sheer fantasy. The unearthing of specific moments of whiteness in action in inner Sydney link also to broader understandings and structures that have replicated and reinforced normative values in Australian society, more generally.

In the next section, I draw together the core themes embedded within *Cities of Whiteness* by revisiting, first, a range of scholarly concepts that have theorized cities. Specifically, I re-consider the ideas of gentrification and urban transformation. Then, I bring theorizing of cities together with thinking about whiteness. Finally, I consider the future of post-colonial geographies of urban whiteness.

Studying Cities

Following contemporary studies of urban transformation in Geography, *Cities of Whiteness* has teased out localized politics that link to broader political economics and identity formations. By tracing some of the specificities of urban transformation/gentrification in inner Sydney, I have unpacked some of the ways that identity has been, and is being, constituted through cultures of production and consumption of 'heritage' housing and new urban 'apartments'. These housing forms have been linked to the (re)production of value-systems that are anchored to very particular histories. Landscapes of heritage housing have become imbued with meaning and nostalgias for bygone eras. Beliefs about heritage and entitlement to place have replicated the colonial project, through reaching back to 'simpler' times – before Indigenous rights – and denying the present, and its unsettling Indigenous presence. In what appears to be a completely different and new genre of housing, Sydney's New York style apartments have elided the troublesome local (rather than the present), by reaching to imaginaries of elsewhere. The interdependence of culture and capital has thus been consolidated, and this has been demonstrated by the ways that they have worked together through processes that perform, consolidate and extend formations of the also largely imagined, but nonetheless forceful 'dominant culture'.

This book has demonstrated a perhaps unexpected, but undeniable link between urban transformation and racialization. In the examples of heritage presented here (in Chapter 3, 'The Good Old Days'), practices of production and consumption have been directly affected by the presence of The Block, and what it represents in a non- or anti-heritage sense for those who live near it. The formations that have defined and constituted 'heritage' have been bound to the historical geographies of colonization, and are therefore highly political. Emergent cultures of denial of, and indifference to, Indigenous (and other) heritages have ensured the ongoing privileging capacities of a highly streamlined form of neo-colonial whiteness and its investments. Expanded consumption options, made available through redevelopment of residential landscapes of many cities around the world, have created cultures that include a range of types of people but with this, various exclusionary capacities have also mobilized. Observation of such formations in inner Sydney has demonstrated that cultures of inclusion (and exclusion) have operated in the service of privileging whiteness. Because of their subtle connections to Anglicized imaginings of elsewhere (temporal, as well as locational) these emergent cultures are more than class or wealth-based elitism. Discourses that promoted 'loft living', for instance, may vary from city to city, but the ends are similar. The commodification of loft living, as discussed in the previous chapter, has been enabled through the production of built environments, and the phantasmagoria of elsewhere. Though conjured from a range of discursive origins, from buildings with Manhattan-referenced names, or images of 'uptown' Manhattan, these various Manhattan dreamings were constructed for a purpose, and that was the sale of real estate in otherwise less than conventionally desirable circumstances.

The production and promotion of loft living in Sydney has appealed to a specific clientele. The new apartment/condominium/loft landscapes provide more than housing; they offer distinction. Only those with the economic surplus to put necessity aside, with the 'urbane' tastes to match, have the capacity to acquire this form of housing. Regardless of the 'ethnicity' of the consumer, the seductive echoes of Manhattan ('New York, New York'), dubbed 'Manhattanisms' by Rem Koolhaas, swirl about within the imaginaries of loft living. These cultural anchors are highly imbued with specific, albeit subtle references to the desirability and even romance of imagined 'whiteness'.

Urban transformation processes have become vehicles for performances of whiteness, and its reification. Whiteness has triumphed as the dominant social ideal, which has been re-anchored to its colonial origins. Although whiteness is expressed through mostly unspoken versions (but, on occasion, highly verbally, as demonstrated in Chapter 3), it appears to be as solid as the sandstone, bricks and mortar of Victorian or Manhattan referenced architecture. In this context, the notion of what constitutes self/us/desirable feeds upon the existence of its opposite: the 'other'/them/the undesirables. In case of inner Sydney, and in Australia (and elsewhere) more generally, the Indigenous 'other' has served this purpose well.

The realities of urban investment in inner Sydney, and in cities generally, mean that urban transformation processes need to be alert to the unpredictable, particularly any threats to profit. The presence of The Block hindered the usual progress of gentrification but astute property developers found ways around the 'problem', through the creation of new urban cultures of escape from the present, and the local. Along with specific heritage designations, apartment developments provided new and flexible approaches to tackling one of Sydney's final gentrification frontiers. The conversions of old industrial sites that were hitherto undercapitalized have proved to be highly effective and profitable beachheads in what were investment 'dead spaces' in the inner city of Sydney (cf Smith 1996). The dramatic re-invention of these spaces occurred through a careful process of cultural encoding. Re-imagining swathes of old industrial spaces has worked to reorient the style and function of much of the inner city in spectacular ways. Encircling and butting up against Aboriginal land, the new incumbents of the transforming city adjust, and are swiftly socialized to the new ways of identifying with (or away from) the urban.

The re-emergence of thinking about urbanisms – as *ways of living* in cities – has largely transcended earlier debates about 'postmodern urbanism'. An observable trend is that developer-driven urban regeneration is often state-sanctioned. However, the concept of a globalized urbanism as an overall city life narrative, and the rise of an associated 'burgeoning geopolitical order' (Dear 2000) of whiteness do not translate completely. In the case of Sydney, local histories have played a substantial role in the ways that whiteness has manifested. The appearance of what seem to be generic expressions of globalized urban whiteness as expressed through contemporary 'loft living', and also in imperial/colonial-referenced heritages, have proved to be highly context dependent.

Racialized Urbanism

Thinking about urbanism in Human Geography is distinct from some other ideas about urbanism that have gone before, or referring to the 'new urbanism' that now inhabits discourses of town planning and architecture. Human geographers, and other urban thinkers, have recently opened the field and found inspiration for thinking about cities by revisiting the writings of urbanists, such as Wirth, Mumford and Simmel, from long ago. In my exploration of urbanism in *Cities of Whiteness*, I have considered the formation of various expressions of urbanity that have accompanied new ways of living in cities. Such urbanisms have been part of the production and consumption of recently conceived residential landscapes. New forms of urbanism have emerged with the usual forms of gentrification, such as those associated with the revitalization of 'heritage' housing landscapes. I have also detailed the rise of exclusionism that has reinforced privilege making.

When high-rise housing in Australia needed a makeover to shake off its second-rate associations, real-estate marketeers rose to the challenge by creating desire for the new landscapes of high-rise housing in old industrial areas and the CBD. The existence and presence of The Block was the obstacle that was overcome in perhaps unexpected ways. It became a consumption opportunity in the process of re-imaging inner Sydney space. It was excluded, through technologies of avoidance, and then – in a somewhat uncanny way – it was included within a mostly unspoken global ghetto script. This has all been a positive outcome for selling real estate, and an opportunity for cosmopolite taste expansion and recognition in the pursuit of distinction. What remains largely absent from the rhetoric of successful urban rejuvenation, through gentrification and apartment development and the rise of associated urbanisms – including the consumption practices of cosmopolitanism – are the material as well as symbolic consequences for the urban Aboriginal community of The Block.

With the idealized Manhattan lifestyles promoted with Sydney's version of 'loft living', the other aspect of Sydney's SoHo Syndrome is the ongoing Harlemization of The Block, which has two seemingly contradictory impacts. The first is that with this de-Indigenization process, 'zero tolerance' seems justifiable, following the US model. The other effect is that this 'ghetto' is simply part of Sydney's maturing cityhood. This ghetto-caricature, along with the zero-tolerance responses, the police sirens and media coverage, are simply consumable parts of Sydney's emergent urbanity that draws on increasingly globalized images of fast, edgy living within the (post)modern cosmopolitan metropolis.

The conceptualizations of urbanism discussed in this book do not fit a 'grand narrative' of urbanism, of singular cultural traits. Local experience tells its own urbanism story, through its historical geographies of place. In Sydney, some of the recently emerged forms of urbanism reflect a trend in consumption practices. For example, the consumption of heritage and old industrial landscapes (now apartments) can be found in many other 'western industrialized' cities. But it is only through acknowledgment of the unique place-based historical geographies that insights to the reproduction and consolidation capacities of whiteness come into view.

Particularities and fantasies of whiteness

Referring to the main conceptual touchstone of this book, I have followed on from studies that go beyond the identification and focus on the 'victims' of racialization processes, to interrogate specific processes in the performance and consolidation of whiteness. As I have posited, and hope to have demonstrated in this book, whiteness is more than an ethnicity, or ethnic identification. It is processes of privileging, and the production of memberships that are not necessarily fixed in time or space, or even to specific bodies. I have traced complex and specific processes of whiteness in a place, and have shown that one main feature of whiteness, and part of its strength, rests in its propensity to be flexible and responsive to context. I hope to have demonstrated that different whiteness processes can operate around separate issues to the same ends, and often without acknowledgement. Whiteness can be a very subtle engagement.

Whiteness is assumed, socialized and institutionalized in Australia. An example that occurred at the very beginning of the research journey that resulted in this book is worthy of mention. I had to gain ethics clearance (from a university), to carry out the first phase of the research. After submitting the research proposal, I was informed that there were special requirements because of the ethics involved in researching Indigeneity. The application I had submitted contained no evidence of consent or approval from the community on The Block. The research had been designed to explore the processes that were occurring in the predominantly non-Aboriginal neighbourhoods that are adjacent to the Aboriginal community, the mass media, government agencies and so on, *about* its presence. After attempts to explain that Indigenous people were not the subject of this research, I finally approached an Aboriginal Elder as directed. He was highly amused that I needed his/their approval and was, at the same time, very pleased to hear that someone wanted to carry out research that was 'well overdue'. Although, like me, he could not see the point of giving his permission for research that had nothing to do with the Aboriginal community directly, he gave it so that I could get on with the work. The other 'communities' were somehow exempt from the vagaries of potential research abuse, or the need for cultural sensitivities. They were not the vulnerable Aboriginal 'other', and this point was not lost on the Elder, who – kindly, and knowingly – gave me permission to *carefully* study 'that other mob'.

The End of (Cities of) Whiteness?

To reiterate one of the overall purposes for writing this book, I have tried to push the notion of whiteness beyond its usual categorizations, and attend to some of the shortcomings of its study. A main contribution to the study of whiteness, which is part of the wider politics of this book, is that I have attempted to address what is, in my view, a spectacular omission. In particular, I have sought to bring Indigeneity (largely ignored by whiteness studies) into contact with the scholarly orbit of the study of whiteness. A caveat to this is that I understand that not all Indigenous peoples around the world are constructed as non-white and along with the benefits such 'passing' (as white) may offer, this has also entrenched another layer of invisibility for some Indigenous peoples. One of these invisibilities has manifested in a shortage of representation of Indigenous peoples within US-dominated whiteness studies. In the Australian context, Aboriginal peoples are rarely constructed as 'white'. When they are, it is often used as a way to denounce their Aboriginal identity, and what is widely perceived to be 'Aboriginal privilege'. It is common to hear sentiments along the lines of 'they don't even look black, how can they claim to be Aboriginal'. Aboriginal peoples have identified as 'black' in a politics that has unified the many Aboriginal clan groups and, in solidarity with the power of African-American civil-rights struggles from the 1960s. But this version of 'black' is not a skin colour – it is an identity that 'comes from within', according to the politics. One overall implication is the categorization of Aboriginal as generally 'other' to white, and this black/white bifurcation can be traced back to colonial/imperial understandings of 'race' (which in the US context rendered Indigenous peoples as 'red'). So in *Cities of Whiteness* I have tried to add to the wider political project of unsettling such categorizations when they work in the service of continuing dispossession. At the same time, I support the politics of solidarity, of otherwise fragmented and dispossessed groups identifying under the same banner, and in this case it is the category 'black'. So through this book I have made a concerted effort to, instead, address the politics of the category 'white' (and by implication, whiteness). In addition, I have advocated the decolonization of the study of 'geographies of whiteness' (Shaw 2006).

To decolonize whiteness I have emphasized two pathways. First, I have attempted to uncouple the fix between whiteness and its too-often-ascribed race/ethnicity, and thereby enable racialized binaries of 'white' and 'black' (as is appropriate in the context of this book, but not fixed to this particular binary) to be disrupted. Second, I have tried to unravel the neo-colonial/imperial contexts within which (the reinforcement of) such binaries produce empowerment. To do this I dipped into a pool of evidence, with cases-in-point and examples provided by events within the rapidly changing inner city of the globalizing metropolis of Sydney.

So, the final agenda of this book is to restate existing disciplinary pleas for a more concerted and ongoing geographical pursuit of (the study of) whiteness. As whiteness studies emerged, largely, from outside the discipline of Geography – with that distinct North American focus – I would like to restate just how well-placed geographers are to engage with locality-based studies that move beyond the geographical constraints of North American research, and its current conceptual binds. In other words, studies of whiteness need to unearth localized processes of racialization and marginalization *within* broader historical geographies of, for example, national and international identity formations (cf Bonnett 2000). I would like also to reiterate a call for a more full engagement with post-colonial/imperial critiques and perspectives for the fresh insights they can provide to the study of whiteness. One consequence of the inclusion of post-colonial perspectives is that a closely related, though geographically distinct, conceptual field has become visible and begs for a more thorough scholarly engagement. Current understandings of what is meant by 'post-colonial' are yet to address, to such depth, the post-*imperial* (Jacobs 1996). For example, racialization at the Anglophone hearts of the UK and the USA (a place that seems to avoid being positioned as either (post)colonial *or* (post) imperial regardless of its settler context), have both largely escaped the theoretical rigour of post-colonial/imperial critiques, and perspectives, until very recently.

Box 5.1 Postcolonial colonizations?

Beyond the main discussions within *Cities of Whiteness*, in other research I have pondered the vexed post-/neo-colonial status of a range of places, particularly around the Pacific, where island states exist in various stages of colonization, decolonization and, in official terms at least, post-colonization. In one research project, I specifically considered the status of colonization for Hawai'i, USA. Hawai'i has multiple identities regarding colonization, and its status as the 50th state of the USA is contested. The occupation of Hawai'i has not only been recognized, President Clinton formally apologized (in 1993), for the overthrow of the Hawaiian Kingdom, which occurred in the late nineteenth century. So under such vexed and multiple conditions, how might Hawai'i USA be subject to a post-colonial/*imperial* critique? Although a story for another time, my preliminary observation (made during fieldwork in December 2003) is that whiteness, in this context too, is not as simple as identifying the racialized segregation of urban spaces. In the case of contested settlement in Honolulu, urban space is trifurcated between Haoles[a] 'Native' Hawaiians (Kanaka 'Oiwi Maoli), and the 'Asian' descendants of the indentured plantation workers, who also make up a significant proportion of the Hawaiian population. Interestingly, a cursory observation of 'heritage' designations was alarmingly familiar. The 'old stuff' deemed worthy enough to be listed on the US 'National Register of Historic Sites', Hawai'i county (www.national-registerofhistoricplaces.com/HI/Hawaii/state2.html), revealed that the heritage of the Kanaka 'Oiwi Maoli appears mainly as pre-invasion archaeology, while non-Kanaka 'Oiwi Maoli 'old stuff', with an emphasis on buildings (and military paraphernalia), holds significance and attracts the scarce preservation dollars. For a Kanaka 'Oiwi Maoli activist I spoke with, 'heritage' is more about the practicalities of life than romanticized artefacts. On the question of heritage, this was his thoughtful response:

> I see – to me there are buildings, there are agricultural sites, there are mountains ... the breeze and everything else, and ... there's a whole heritage here. Of course, I cannot see what you see. I cannot see anything specifically about your ancestry, but I ... see just what I can see. So I have a material kind of heritage experience, which is probably not what you're having. I see ... bringing in the groups of people who need to touch base with their own identity for political reasons, apart from anything else, political, personal empowerment reasons. (Eric Enos, Ka'ala Farm, December 2003)

Box 5.1 (Continued)

The task of maintaining post-colonial/imperial perspectives requires vigilance, particularly to the colonial histories and legacies that have given context to, and provided the foundations for – in this example – designations of 'heritage'. Such designations resist relegation to the status of 'archaeology', and focus more on (the ravages of) everyday living.

[a] The term 'Haole', which usually refers to non-Kanaka 'Oiwi Maoli (Kanaka 'Oiwi Maoli is a unifying term for 'native' Hawaiians), is not as specifically applied as 'gubba' is in Australia, but the meanings are similar. See http://en.wikipedia.org/wiki/Haole for more on popular understandings.

My underlying request in *Cities of Whiteness* is that the disciplinary of geography engage more fully or even *re*-engage with issues of urban racialization, and in particular, the somewhat overlooked, or avoided, arena of whiteness. I have attempted to demonstrate that one of the strengths of whiteness is its capacity to encompass any power formation, be it economic capacity or political clout. This capacity to transcend ethnicity has enabled whiteness to hide behind a blanket of *apparent* inclusiveness, of 'ethnic' diversity. Memberships have been open but only to those with the various capacities to join in the separation, or disempowerment, of the 'other' (even if only temporarily). Whether or not this book fits within the categorization of the sub-, or quasi- discipline of whiteness studies, is open to speculation. If it must be categorized, it should be considered as part of contemporary human geographies that have been influenced by critical analyses of cultures, and cultural production. In my examples, whiteness has performed as a cultural formation. It is part social ideal, part normative production, and part embodiment of privilege. Following Bonnett's (2000, 134) observation, then, this book has attempted to document some of the 'plural constitutions and multiply lived experiences of whiteness'.

As Ghassan Hage (1998) also astutely observed, whiteness – and its benefits – is an aspiration, which can be pursued regardless of one's ethnicity, or how one is identified by others. As this book demonstrates, through a process of continual (re)positioning and privileging, one of the processes of whiteness is its capacity to ascribe racialized 'encodings'. *Cities of Whiteness* has provided extensive examples of such racialized encoding as they have manifested through mass-media portrayals (of the racialized 'other'), policing and operated in educational facilities. The overall and ongoing representation of such institutions has been that they are unethnicized or ethnically neutral and, as such, represent the status quo regardless of who belongs to these groups. In other words, the operations of whiteness, just like its definition, need to be fully unpacked. As with 'capitalism' (cf. Gibson-Graham 1996), the overarching understanding that whiteness is an ethnicity, or 'race', needs to be de-essentialisied, and overdetermined. Within the domain of geographical scholarship, if we are to accept that whiteness (like blackness) is a racialized/ethnicized construction then closer examination, rather than avoidance, is also a useful disciplinary trajectory. Whiteness is discursively active regardless of its (in)visibility within societies or, for that matter, within a self-reflexive scholarly discipline.

In Australia, the non-Aboriginal majority (and some Indigenous people as well), including many of those who walked *en masse* across the Sydney Harbour Bridge in support of reconciliation in 2000,[8] still regard the urban Aboriginal community of The Block as a disgrace, a thorn in the side of nationhood. It is simply easier to blame Indigeneity for the state of The Block than to confront the issues it manifests so abjectly. Generally, the voices that resist whiteness are marginalized, sometimes even by their (our) own unconscious engagements with the subtleties of whiteness processes. It is clear that a more pragmatic way of dealing with the difficulties of cohabitation might include de-fantasizing whiteness from its sometimes romantic but often hostile and resentful ideas about Indigenous people. Socio-cultural shifts that have accompanied such rapid urban development have consolidated whiteness through its expressions of (select) heritage sensibilities and new urbanities. These have occurred to the detriment of what was originally and hopefully conceived as the 'black capital' of Australia.

As bulldozers demolish more precious buildings on The Block, and larger forces are set to pose another set of challenges to its existence, Aboriginal Elders have simply reiterated the significance of their place in the city, which will always be a meeting place for dispossessed Aboriginal peoples. Where houses once stood, there are now wide-open spaces dotted with tents and campfires that seem to indicate a permanent form of housing. This space is earmarked, yet again, for redevelopment by forces external to Aboriginal community, the shape of which remains uncertain. For the moment, however, and against the ever-increasing odds and regardless of its state, this tiny Aboriginal place continues to survive at the heart of a rapidly transforming and increasingly global metropolis.

Notes

1 External forces have fuelled the surge in property values with the change in Sydney's status to a (quasi) global city over the last decade.
2 The Redfern-Waterloo Authority's development plan for Redfern-Waterloo has included 'a $10 million community health centre at the former Redfern Courthouse' (*South Sydney Herald*, 1, 44, 2006).
3 ABC News Online http://www.abc.net.au/news/Indigenous/ab-25oct2000-10.htm
4 The Australian Heritage Commission was replaced by the Australian Heritage Council in 2003.
5 See Chapter 1 for a discussion of the '[s]ickening ['race'] Riots' that 'Rocked Sydney' (*Sydney Morning Herald*, 16 February 2004, 1), and the aftermath.
6 Go to www.ahc.org.au/redevelop/redevelop.html for details.
7 'Pemulwuy lives, say hundreds who attend Block protest', *South Sydney Herald*, 1, 44 (2006).
8 Reconciliation is a fraught and debated term. For some it suggests bringing together two 'sides' without necessarily acknowledging the power relations of colonialism. The vision of the Council for Aboriginal Reconciliation is 'a united Australia that respects this land of ours, values the Aboriginal and Torres Strait Islander Heritage, and provides justice and equity for all' (Draft Declaration for Reconciliation, Council for Reconciliation).

Bibliography

Allen T. 1994. *The Invention of the White Race. Volume One: Racial Oppression and Social Control*. London: Verso

Anderson, K. 1993a. Constructing geographies: 'race', place and the making of Sydney's Aboriginal Redfern, in P. Jackson and J. Penrose (eds), *Constructions of Race, Place and Nation*. London: UCL Press

Anderson K. 1993b. Place narratives and the origins of the Aboriginal settlement in inner Sydney, 1972–73. *Journal of Historical Geography* 19, 3, 314–335

Anderson K. 1998. Sites of difference: beyond a cultural politics of race polarity, in R. Fincher and J.M. Jacobs (eds), *Cities of Difference*. New York: Guilford Press

Anderson K. 1999. Reflections on Redfern, in E. Stratford (ed.), *Australian Cultural Geographies*. Melbourne: Oxford University Press

Anderson K. 2000. 'The beast within': race, humanity, and animality. *Environment and Planning D: Society and Space* 18, 301–320

Anderson, K., and Gale, F. (eds) 1992. *Inventing Places: Studies in Cultural Geography*. Longman Cheshire, Melbourne

Anon. 1974. Aboriginal Australia - The Redfern Example, *Aboriginal News* 1, 9, October, 16

Armstrong H. 1994. Cultural continuity in multicultural sub/urban places, in K. Gibson and S. Watson (eds), *Metropolis Now*. Australia: Pluto

Ashworth G. and Tunbridge J. 1990. *The Tourist Historic City*. London: Belhaven Press

Atkinson R. and Bridge G. 2005. *The New Urban Colonialism: Gentrification in a global context*. London Routledge.

Attwood B. 1989. *The Making of the Aborigines*. Sydney: Allen and Unwin

Attwood B. and Arnold J. (eds) 1992. *Power, knowledge and Aborigines*. Bundoora: La Trobe University Press

Back L. 2002. Aryans reading Adorno: cyber-culture and twenty-first-century racism. *Ethnic and Racial Studies* 24, 628–652

Badcock B. 1996. 'Looking-glass' views of the city. *Progress in Human Geography* 20, 1, 91–99

Barnes T. and Duncan J. 1992. *Writing Worlds: Discourse, Text and Metaphor in the Representation of Landscape.* London and New York: Routledge

Barthes R. 1986. Semiology and the urban, in M. Gottdiener and A. Ph. Lagopoulos, (eds), *The City and the Sign.* Columbia University Press, New York

Beauregard R. 1990. Trajectories of neigborhood change: the case of gentrification. *Environment and Planning A* 22, 855–874

Beauregard R. 1993. *Voices of Decline: The Postwar Fate of US Cities.* Oxford: Basil Blackwell

Beauregard R. 1999. Break dancing on Santa Monica Boulevard. *Urban Geography* 20, 5, 396–399.

Beckett A. 1994. The safe way to shop. *Good Weekend* (Sydney Morning Herald), 7 May

Bennett T. 1993. History on the rocks, in J. Frow and M. Morris (eds), *Australian Cultural Studies: A Reader.* Sydney: Allen and Unwin

Benjamin W. 1978 [1935]. Paris: capital of the nineteenth century, in P. Kasinitz (ed.), *Metropolis: Centre and Symbol of Our Times.* New York: New York University Press

Bhabha H.K. (ed.) 1990. *Narration and Nation.* London and New York: Routledge

Bhabha H.K. 1994. *The Location of Culture.* London: Routledge

Bhabha H.K. 1998. The white stuff. *Artforum* 36, 9, 21–24

Blaut J.M. 1992. The theory of cultural racism. *Antipode* 24, 4, 289–299

Bondi L. 1991. Gender divisions and gentrification: a critique. *Transactions, Institute of British Geographers* 16, 290–298

Bonnett A. 1992. Anti-racism in 'white' areas: the example of Tyneside'. *Antipode* 24, 1, 1–15

Bonnett A. 1993. Contours of crisis: anti-racism and reflexivity, in P. Jackson and J. Penrose (eds), *Constructions of Race, Place and Nation.* London: UCL Press

Bonnett A. 1996. Anti-racism and the critique of 'white' identities. *New Community* 22, 1, 97–110

Bonnett A. 1997. Geography, 'race' and Whiteness: invisible traditions and current challenges. *Area* 29, 3, 193–199

Bonnett A. 2000. *White Identities: Historical and International Perspectives.* Harlow: Prentice Hall

Bonnett A. 2002. The Metropolis and White Modernity. *Ethnicities* 2, 3, 349–366

Bonyhady T. 1995. The battle for Balmain, in P. Troy, *Australian Cities: Issues, Strategies and Policies for Urban Australia in the 1990s.* Cambridge: Cambridge University Press

Bourdieu P. 1984. *Distinction: A Social Critique of the Judgement of Taste.* London: Routlege

Bowman A. 1995. Reshaping the city, in J. Walter, H. Hinsley and P. Spearritt, *Changing Cities: Reflections on Britain and Australia.* London: Sir Robert Menzies Institute of Australian Studies

Boyer M.C. 1998. *The City of Collective Memory: Its Historical Imagery and Architectural Entertainments.* Massachusetts: MIT Press

Brockie, J. 1991. *Cop It Sweet.* Sydney: ABC TV Documentaries

Brown W. 1995. *States of Injury: Power and Freedom in Late Modernity.* Princeton, New Jersey: Princeton University Press

Bulbeck C. 1993. *Social Sciences in Australia: An Introduction.* Sydney: Harcourt Brace Janovich

Burgmann M. and Burgmann V. 1998. *Green Bans, Red Union: Environmental Aactivism and the New South Wales Builders Labourers' Federation.* Sydney: UNSW Press

Burgmann, V. 1993. *Power and Protest: Movements for Change in Australian Society.* Sydney: Allen and Unwin

Burnley I.H. (ed.) 1974. *Urbanization in Australia: the Post-war Experience.* Cambridge: Cambridge University Press

Burnley I. and Murphy P. 1994. *Immigration, Housing Costs and Population Dynamics in Sydney.* Canberra: Australian Govt. Public Service

Caldeira T.P.R. 1996. Fortified enclaves: the new urban segregation. *Public Culture* 8, 303–328

Cameron L. and Craig M. 1985. A decade of change in inner Sydney. *Urban Policy and Research* 3, 4 22–30

Castells M. 1977. *The Urban Question: A Marxist Approach.* London: Edward Arnold

Castells M. 1983. *The City and the Grass Roots.* London: Edward Arnold

Castles S., Kalantzis M., Cope B. and Morrissey, M. 1988. *Mistaken Identity: Multiculturalism and the Demise of Nationalism in Australia.* Sydney: Pluto

N. 2005. The epistemology of particulars. *Geoforum* 36, 5, 541–44

Clark D. 2000. World urban development: processes and patterns at the end of the Twentieth Century. *Geography* 85, 1, 15–23

Cole D. 1987. Artists and urban redevelopment. *Geographical Review* 77, 4, 391–407

Collins J. 1995. *Architecture of Excess: Cultural Life in the Information Age.* New York: Routledge

Conybeare Morrison and Partners 1990. *The University of Sydney Strategy Plan.* University of Sydney

Coombs A. 1996. *Sex and Anarchy: The Life and Death of the Sydney Push.* Victoria: Penguin

Costello L.N. and Dunn K.M. 1994. Resident action groups in Sydney: people power or rat-bags? *Australian Geographer* 25, 1, 61–76

Costello L.N. and Hodge S. 1999. Queer/clear/here: destabilising sexualities and spaces, in E. Stratford (ed.), *Australian Cultural Geographies*. Sydney: Meridian

Crang M. 1994. On the heritage trail: maps of and journeys to olde Englande. *Environment and Planning D: Society and Space* 12, 341–355

Crang M. 1998. *Cultural Geography*. London: Routledge

Cresswell T. 1996. *In Place, Out of Place: Geography, Ideology and Transgression*. Minneapolis: University of Minnesota Press

Crilley D. 1993. Architecture as advertising: constructing the image of redevelopment, in G. Kearns and C. Philo (eds), *Selling Places: The City as Cultural Capital, Past and Present*. Oxford: Pergamon Press

Cunneen C. 1990. *Aboriginal-Police Relations in Redfern: With Special Reference to the 'Police Raid' of 8 February 1990*. Report Commissioned by the National Inquiry into Racist Violence. Human Rights and Equal Opportunity Commission, May

Curson P. 1985. The impact of inequality: the Sydney plague epidemic of 1902, in I. Burnley and J. Forrest (eds), *Living in Cities: Urbanisation and Society in Metropolitan Australia*. Sydney: Allen and Unwin

Curthoys A. 2000. 'An uneasy conversation: the indigenous and the multicultural', in John Docker and Gerhard Fischer (eds), *Race Colour and Identity in Australia and New Zealand*. Sydney: UNSW Press.

Dalby S. and Mackenzie F. 1997. Reconceptualising local community: environment, identity and threat. *Area* 29, 2, 99–108

Daly M. 1992. *Sydney Boom, Sydney Bust*. Sydney: Allen and Unwin

Daly M. 1998. Reshaping Sydney: the inner city revival. *Urban Policy and Research* 16, 1, 59–63

Davis M. 1990. Fortress Los Angeles: the militarisation of urban space, in M. Sorkin, *Variations on a Theme Park*. New York: Hill and Wand

Davis M. 1998. *Ecology of Fear: Los Angeles and the Imagination of Disaster*. New York: Metropolitan Books

Dear M. 1992. Understanding and overcoming the NIMBY syndrome. *Journal of the American Planning Association* 58, 3, Summer, 288–381

Dear M. 2000. *The Postmodern Urban Condition*. Oxford: Blackwell

Dear M. and Flusty S. 1998. Postmodern urbanism. *Annals of the Association of American Geographers* 88, 1, 50–72.

Dear M. and Takahashi L. 1997. The changing dynamics of community opposition to human service facilities. *Journal of the American Planning Association* 63, 1, 79–93

Dear M. and Taylor S.M. 1982. *Not on Our Street: Community Attitudes Toward Mental Health Care*. London: Pion

de Certeau M. 1984. *The Practice of Everyday Life*. Berkeley: University of California Press

Douglas M. 1966. *Purity and Danger*. London: Routledge

Driver F. 1995. Visualising geography: A journey to the heart of the discipline. *Progress in Human Geography* 19, 123–134.

DuBois W.E.B. 1899. *The Philadelphia Negro*. Philadelphia: University of Pennsylvania Press

Duncan J.S. 1980. The superorganic in American cultural geography, *Annals, Association of American Geographers* 70, 181–98

Dunn K.M. 1993. The Vietnamese concentration in Cabramatta: site of avoidance and deprivation, or island of adjustment and participation? *Australian Geographical Studies* 31, 2, 228–245

Dunn K.M. 1998. Rethinking ethnic concentration: the case of Cabramatta, Sydney. *Urban Studies* 35, 3, 503–527

Dwyer O.J. and Jones J.P. 2000. White socio-spatial epistemology. *Social and Cultural Geography* 1, 209–222

Dyer R. 1988. White. *Screen* 28, 4, 44–64

Dyer R. 1997. *White*. London: Routledge

Eade J. (ed.) 1997. *Living in the Global City*. London: Routledge

Engels B. 1994. Capital flows, redlining and gentrification: the pattern of mortgage lending and social change in Glebe, Sydney, 1960–1984. *International Journal of Urban and Regional Research* 18, 4, 28–58

Engels B. 1999. Property ownership, tenure and displacement: in search of the process of gentrification. *Environment and Planning A* 31, 1473–1495

Ezekiel R. 1995. *The Racist Mind: Portraits of American Neo-Nazis and Klansmen*. New York: Viking

Fainstein S. 2000. New directions in planning theory. *Urban Affairs Review* 35, 4, 451–478

Featherstone M. (ed.) 1990. *Global Culture*. London: Sage

Fielder J. 1991. Purity and pollution: Goonininup/The Old Swan Brewery. *Southern Review* 24, 1, 34–42

Fielder J. 1995. Sacred sites and the city: urban Aboriginality, ambivalence, and modernity, in R. Wilson and A. Dirlik (eds), *Asia/Pacific as Space of Cultural Production*. Durham, New Carolina: Duke University Press

Fincher R. 1987. Defining and explaining urban social movements. *Urban Geography* 8, 2, 152–160

Fincher R. and Jacobs J.M. (eds) 1998. *Cities of Difference*. New York: Guilford Press

Fitzgerald S. 1987. *Rising Damp: Sydney 1870–90.* Melbourne: Oxford University Press

Fitzgerald S. and Keating C. 1991. *Millers Point: The Urban Village*. Sydney: Hale and Iremonger

Forrest J. and Burnley I. (eds) 1985. *Living in Cities: Urbanism and Society in Metropolitan Australia*. Sydney: Allen and Unwin

Foucault M. 1977. *Discipline and Punish: The Birth of the Prison*. London: Penguin

Frankenberg R. 1993. *The Social Construction of Whiteness: White Women, Race Matters*. Minneapolis: University of Minnesota

Fredrickson G. 1981. *White Supremacy: A Comparative Study in American and South African History*. New York: Oxford University Press

Frost L. 1992. Suburbia and inner cities, in A. Rutherford (ed.), *Populous Place: Australian Cities and Towns*. Sydney: Dangaroo Press

Gabriel J. 1998. *Whitewash: Racialized Politics and the Media*. London: Routledge

Gale F. with Brookman A. 1972. *Urban Aborigines*. Canberra: Australian National University Press

Gandhi L. 1998. *Postcolonial Theory: A Critical Introduction*. Sydney: Allen and Unwin

Geertz C. 1973. *The Interpretation of Cultures*. New York: Basic Books

Gelder K. and Jacobs J.M. 1998. *Uncanny Australia: Sacredness and Identity in a Postcolonial Nation*. Melbourne University Press

Gibson C. and Pagan R. 2000. Rave cultures in Sydney, Australia: mapping youth spaces in media discourses, in M.A. Wright (ed.), *Dance Culture, Party Politics and Beyond*. London: Verso

Gibson-Graham J.K. 1996. *The End of Capitalism (As We Knew It): A Feminist Critique of Political Economy*. Oxford: Blackwell

Goodwin M. 1993. The city as commodity: the contested spaces of urban development, in G. Kearns and C. Philo (eds), *Selling Places: The City as Cultural Capital Past and Present*. Oxford: Pergamon

Goss J. 1997. Representing and re-presenting the contemporary city: Progress Report. *Urban Geography* 18, 2, 180–188

Graham B., Ashworth G.J. and Tunbridge J.E. 2000. *A Geography of Heritage: Power, Culture and Econom.*, London: Arnold

Gregory S. 1993. Race, rubbish, and resistance: empowering difference in community politics. *Cultural Anthropology* 8, 1, 24–48

Gunew S. 1992. Against multiculturalism: Rhetorical images. *Typereader, Journal of the Centre for Studies in Literary Education* 7, Autumn, 28–41

Hage G. 1993. Anglo-Celtics today: cosmo-multiculturalism and the phase of the fading phallus. *Communal/Plural* 4, University of Western Sydney

Hage G. 1997. At home in the entrails of the West: multiculturalism, 'ethnic food' and migrant home building, in H. Grace, G. Hage, L. Johnson, J. Langsworth and M. Symonds (eds), *Home/World: Space, Community and Marginality in Sydney's West*. Sydney: Pluto

Hage G. 1998. *White Nation: Fantasies of White Supremacy in a Multicultural Society*. Sydney: Pluto

Harvey D. 1972. Society, the city and the space-economy of urbanism. Association of American Geographers Resource Paper, Washington DC

Harvey D. 1973. *Social Justice and the City*. London: Edward Arnold

Harvey D. 1989. *The Condition of Postmodernity*. Oxford: Blackwell

Harvey D. 1990. Between space and time: reflections on the geographical imagination. *Annals, Association of American Geographers* 80, 418–434

Health Department New South Wales 1999. *Community Consultation on Drug and Alcohol Services in Redfern,* Report no. 99455RP1

Hewison R. 1987. *The Heritage Industry: Britain in a Climate of Decline.* London: Methuen

Hickman, M.J. and Walter, B. 1997. *Discrimination and the Irish Community in Britain.* CRE

Hillier J. 1996. The gaze in the city: video surveillance in Perth. *Australian Geographical Studies* 34, 1

Hillier J. and McManus P. 1994. Pull up the drawbridge: fortress mentality in the suburbs, in K. Gibson and S. Watson (eds), *Metropolis Now.* Sydney: Pluto

Hobsbawn E. and Ranger T. (eds) 1983. *The Invention of Tradition.* Cambridge: Cambridge University Press

hooks b. 1992. Representing whiteness in the black imagination, in L. Crossberg, C. Nelson and P.A. Treichler (eds), *Cultural Studies.* London and New York: Routledge.

Horvath R. and Engels B. 1985. The residential restructuring of inner Sydney, in I. Burnley and J. Forrest (eds), *Living In Cities: Urbanism and Society in Metropolitan Australia.* Sydney: Allen and Unwin

Howe R. 1994. Inner suburbs: from slums to gentrification, in L. Johnson (ed.), *Suburban Dreaming.* Victoria: Deakin University Press

Ignatiev N. 1995. *How the Irish Became White.* London: Routledge

Jackson E. 1997. *The Block.* Australian Broadcasting Corporation Television Documentaries, Sydney, 12 May

Jackson P. 1985a. Social geography: race and racism. *Progress in Human Geography* 9, 1, 99–108

Jackson P. 1985b. Urban ethnography. *Progress in Human Geography* 9, 1, 157–76

Jackson P. (ed.) 1987. *Race and Racism: Essays in Social Geography.* London: Allen and Unwin

Jackson P. 1989. *Maps of Meaning: an Introduction to Cultural Geography.* London: Unwin Hyman

Jackson P. 1991a. Mapping meanings: a cultural critique of locality studies. *Environment and Planning A* 23, 215–228

Jackson P. 1991b. The crisis of representation and the politics of position (guest editorial). *Environment and Planning D: Society and Space* 9, 131–134

Jackson P. 1995. Manufacturing meanings: culture, capital and urban change, in A. Rogers and S. Vertovec (eds), *The Urban Context: Ethnicity, Social Networks and Situational Analysis.* Oxford: Berg

Jackson P. 1999a. Postmodern urbanism and the ethnographic void. *Urban Geography* 20, 4, 400–402.

Jackson P. 1999b. Commodity cultures: the traffic in things. *Transactions of the Institute of British Geographers* 24, 95–108

Jackson P. and Jacobs J.M. 1996. Postcolonialism and the politics of race. Editorial, *Environment and Planning D: Society and Space* 14, 1–3

Jacobs J. M. 1992. Cultures of the past and urban transformation: the Spitalfields Market Redevelopment in East London, in K. Anderson and F. Gale (eds), *Inventing Places: Studies In Cultural Geography.* Melbourne: Longman Cheshire

Jacobs J.M. 1993. The city unbound: qualitative approaches to the city. *Urban Studies* 30, 4/5, 827–848

Jacobs J.M. 1996. *Edge of empire: Postcolonialism and the City.* London: Routledge

Jacobson M.F. 1998. *Whiteness of a Different Color: European Immigrants and the Alchemy of Race.* Cambridge, Massachusetts: Harvard University Press

Jager M. 1986. Class definition and the aesthetics of gentrification: Victoriana in Melbourne, in N. Smith and P. Williams (eds), *Gentrification of the City.* Australia: Allen and Unwin

Jakubowicz A. 1984. The green ban movement: urban struggle and class politics, in J. Halligan and C. Paris (eds), *Australian Urban Politics.* Melbourne: Longman Cheshire

Jenks C. 1981. *The Language of Post-Modern Architecture.* New York: Rozzoli

Jones R. 1997. Sacred sites or profane buildings? Reflections on the Old Swan Brewery conflict in Perth, Western Australia, in B.J. Shaw and R. Jones (eds), *Contested Urban Heritage: Voices from the Periphery.* Sydney: Ashgate

Kasinitz P. (ed.) 1995. *Metropolis: Centre and Symbol of Our Times.* New York: New York University Press

Keating C. 1991. *Surry Hills.* Sydney: Hale and Iremonger

Kearns G. and Philo C. (eds) 1993. *Selling Places: The City as Cultural Capital, Past and Present.* Oxford: Pergamon Press

Keith M. 1993. From punishment to discipline? Racism, racialization and the policing of Social Control, M. Cross and M. Keith (eds), *Racism, The City and The State.* London and New York: Routledge

Keith M. 1996. Street sensibility: negotiating the political by articulating the spatial, in A. Merrifield and E. Swyngedouw (eds), *The Urbanisation of Injustice.* London: Lawrence and Wishart

Kendig H. 1979. *New Life for Old Suburbs: Post-war Land Use and Housing in the Australian City.* Sydney: Allen and Unwin

Kimmel M. 2000. 'White men are this nation': right-wing militias and the restoration of rural American masculinity. *Rural Sociology* 65, 582–605

King A.D. 1992. Rethinking colonialism: an epilogue, in N. Alsayyad (ed.), *Forms of Dominance: On the Architecture and Urbanism of the Colonial Enterprise.* Aldershot: Avebury

Kohen J.L. 2000. First and last people: Aboriginal Sydney, in J. Connell (ed.), *Sydney: The Emergence of a World City.* Oxford University Press

Koolhaas R. 1978. *Delirious New York: A Retroactive Manifesto for Manhattan.* New York: Oxford University Press

Kristeva J. 1982. *Powers of Horror.* New York: Columbia University Press

Kristeva, J. 1991. (transl. by L.S.Roudiez) *Strangers to Ourselves.* New York: Columbia University Press

Lake R.W. 1999. Postmodern urbanism? *Urban Geography* 20, 5, 393–395

Lambert D. 2001. Liminal figures: poor whites, freedmen, and racial re-inscription in colonial Barbados. *Environment and Planning D: Society and Space* 19, 3, 335–350

Latham A. 1999. Powers of engagement: on being engaged, being different, and urban life. *Area* 31, 1

Lawrence B. 2004. *'Real' Indians and Others: Mixed-Blood Urban Native Peoples and Indigenous Nationhood.* Vancouver: UBC Press

Lauria M. and Knopp L. 1985. Towards an analysis of the role of gay communities in the urban renaissance. *Urban Geographer* 6, 152–169

Lees B.G and Shaw W.S. 2002. GISc and cultural applications: social polarization and spatial co-incidence. Paper presented at the Association of American Geographers annual meeting in Los Angeles.

Lees L. 1994. Rethinking gentrification: beyond the position of economics and culture. *Progress in Human Geography* 18, 2, 137–150

Lees L. 1996. In the pursuit of difference: representations of gentrification. *Environment and Planning A* 28, 453 – 470

Lees L. 1999. The weaving of gentrification discourse and the boundaries of the gentrification community (guest editorial). *Environment and Planning D: Society and Space* 17, 127–132

Lees L. 2000. A reappraisal of gentrification: towards a 'geography of gentrification'. *Progress in Human Geography* 24, 3, 389–408

Lefebvre H. 1971. *Everyday Life in the Modern World.* New York: Harper and Row

Leitner H. 1992. Urban geography: responding to new challenges. *Progress in Human Geography* 16, 1, 105–118

Ley D. 1974. *The Black Inner City as Frontier Outpost.* Washington DC: Association of American Geographers, Monograph 7

Ley D. 1980. Liberal ideology and the postindustrial city. *Canadian Geographer* 25, 238–258

Ley D. 1986. Alternative explanations for inner-city gentrification: a Canadian assessment. *Annals of the Association of American Geographers* 76, 521–535

Ley D. 2004. Transnational spaces and everyday lives. *Transactions of the Institute of British Geographers* 29, program 151–64.

Lowenthal D. 1985. *The Past Is a Foreign Country.* Cambridge: Cambridge University Press

Lozanovksa M. 1994. Abjection and architecture: the migrant house in multicultural Australia, in L. Johnson (ed.), *Suburban Dreaming*, Victoria: Deakin University Press

McCarthy, C., Rodriguez, A., Meecham, S., et al. 1997. Race, suburban resentment, and the representation of the inner city in contemporary film and television, in M. Fine, L. Weis, L.C. Powell and L.M. Wong (eds), *Off White: Readings on Race, Power, and Society*. New York: Routledge

McInnes, P. 1967. Commentary, in P. Troy (ed.), *Urban redevelopment in Australia*. Canberra: Australian National University.

Macintyre S. & Clark A. 2003. *The History Wars*. Melbourne: Melbourne University Publishing

McKay B. (ed.) 1999. *Unmasking Whiteness: Race Relations and Reconciliation*. Queensland: Queensland Studies Centre.

McKenzie E. 1993. Trouble in Privatopia. *The Progressive* 57, 10, 30

Mani L. 1990. Multiple mediations: feminist scholarship in the age of multinational reception. *Feminist Review* 35, Summer 24–41

Massey D. and Allen J. (eds) 1984. *Geography Matters! A Reader*. Cambridge: Open University/Cambridge University Press

Merrifield A. 1996. Public space: integration and exclusion in urban life. *City: Analysis of Urban Trends, Culture, Theory, Policy, Action* 5–6, November

Mickler S. 1991. The battle for Goonininup. *Arena* 96, 69–88.

Mickler S. 1998. *The Myth of Privilege: Aboriginal Status, Media Visions, Public Ideas*. Fremantle, Western Australia: Fremantle Arts Centre Press

Milicevic A S. 2001. Radical intellectuals: what happened to the new urban sociology? *International Journal of Urban and Regional Research* 25, 4, 759–783

Mills C. 1988. 'Life on the Upslope': the postmodern landscape of gentrification. *Environment and Planning D: Society and Space* 6, 169–189

Mills C. 1993. Myths and meanings of gentrification, in J. Duncan and D. Ley (eds), *Place, Culture, Representation*. London: Routledge

Mitchell J.C. 1983. Case and situational analysis. *Sociological Review* 31, 2, 187–211

Morgan G. 1994. Acts of enclosure: crime and defensible space in contemporary cities, in K. Gibson and S. Watson (eds), *'Metropolis Now': Crime and Defensible Space in Contemporary Cities*. Sydney: Pluto

Morrill R. 1965. The Negro ghetto: problems and alternatives', *Progress in Human Geography* 17, 3, 1993, 349–353

Morrison T. 1970. *The Bluest Eye*. London: Vintage

Morrison T. 1992. *Playing in the Dark: Whiteness and the literary imagination*. Cambridge, Massachusetts: Harvard University Press

Morrison T. 1998. *Paradise*. New York: Alfred A. Knopf

Muecke S. 1992. *Textual Spaces: Aboriginality and Cultural Studies*. Sydney: University of NSW Press

Mumford 1995 [1938]. The culture of cities, in Kasnitz P. (ed.), *Metropolis: Centre and Symbol of Our Times*. New York: New York University Press

Mundey J. 1981. *Green Bans and Beyond*. Sydney: Angus and Robertson

Murphy P. and Watson S. 1997. *Surface City: Sydney at the Millennium*. Sydney: Pluto

Nash C. 2003. Cultural geography: anti-racist geographies. *Progress in Human Geography* 27, 637–648

Newman O. 1972. *Defensible Space: Crime Prevention through Urban Design*. New York: Macmillan & Co.

North M. and Christie S. 1997. *The De-industrialisation of Chippendale: Industrial Heritage in the Post-Industrial City*. Sydney: Department of Urban Affairs and Planning, Heritage Assistance Plan

Oliver L.J. 2002. The making and unmaking of Whiteness/Out of Whiteness: color, politics and culture. *Callaloo* 25, 1272–1278

Peach C. (ed.) 1975. *Urban Social Segregation*. London: Longman

Peach C. 1993. Commentary 2: classics in human geography revisited. *Progress in Human Geography* 17, 3, 350–351

Pendall R. 1999. Opposition to housing: NIMBY and beyond. *Urban Affairs Review* 35, 1, 112–136

Peters E. 1998. Subversive spaces: First Nations women and the city. *Environment and Planning D: Society and Space* 16, 665–685

Peters E. 2005. Review of indigeneity and marginalisation: planning for and with urban Aboriginal communities in Canada. *Progress in Planning* 63, 327–404

Plotkin S. 1990. Enclave consciousness and neighborhood activism, in J.M. Kling and P.S. Posner (eds), *Dilemmas of Activism: Class, Community, and Politics of Local Mobilization*. Philadelphia: Temple University Press

Podmore J. 1998. (Re)Reading the 'loft living' *habitus* in Montreal's inner city. *International Journal of Urban and Regional Research* 22, 2, 283–302

Powell A.J. 1967. 76 in Sydney: Redevelopment in a metropolitan region context, in P. Troy (ed.), *Urban Redevelopment in Australia*. Canberra: Urban Research Unit, Australian National University

Pulido L. 2002. Reflections on a White Discipline. *Professional Geographer* 54, 42–49

Rapport A. 1995. Review of Herzfeld. *Man* 1, 3, September, 645–646

Redfern P. 1997. A new look at gentrification: 1, gentrification and domestic technologies. *Environment and Planning A* 29, 1275–1296

Reynolds H. 1996. *Frontier: Reports From the Edge of White Settlement*. Sydney: Allen and Unwin

Reynolds H. 1999. *Why Were We Never Told?* Victoria: Penguin

Robinson J. 1994. White women researching/representing 'Others': from anti-apartheid to postcolonialism?, in A. Blunt and G. Rose (eds), *Writing Women and Space: Colonial and Postcolonial Geographies*. New York: Guilford Press

Roddewigg R. 1978. *Green Bans: The Birth of Australian Environmental Politics: A Study in Public Opinion and Participation.* Sydney: Hale and Iremonger

Roediger D. 1991. *The Wages of Whiteness: Race and the Making of the American Working Class.* London: Verso

Rosaldo R. 1989. Imperial nostalgia. *Representations,* 26, Spring, 107–121

Rose H., Peach C. and Morrill R. 1993. Classics in human geography revisited: Morrill R. 1965, The Negro Ghetto: problems and alternatives. *Progress in Human Geography* 17, 3, 349–353

Roseth J. 1981. Residential conversions in Sydney. *RAPIJ,* May, 76–78

Said E, 1978. *Orientalism: Western Conceptions of the Orient,* London: Penguin

Sassen S. 1991. *The Global City.* New Jersey, Princeton University Press

Sassen-Koob S. 1985. *Issues of Core and Periphery: Labour Migrations and the New International Division of Labour.* Hong Kong: International Sociological Association Conference on Urban and Regional Impacts of New International Division of Labour, University of Hong Kong

Sayer A. 1989. The 'new' regional geography and problems of narrative. *Environment and Planning D: Society and Space* 7, 253–276

Sayer A. 1993. Commentary. *Progress in Human Geography* 17, 509–551

Sennett R. 1976. Community becomes uncivilised, in Kasinitz P. (ed.), *Metropolis: Centre and Symbol of Our Times.* New York: New York University Press

Sennett R. 1990. *The Conscience of the Eye: The Design and Social Life of Cities.* London: Faber and Faber

Sennett R. 1994. *Flesh and Stone: The Body and the City in Western Civilization.* London: Faber and Faber

Seth S., Gandhi L. and Dutton M. 1998. Postcolonial Studies: a beginning. *Postcolonial Studies* 1, 1, 7–11

Shaw B.J. and Jones R. (eds) 1997. *Contested Urban Heritage: Voices from the Periphery.* Sydney: Ashgate.

Shaw W.S. 2000. Ways of whiteness: Harlemising Sydney's Aboriginal Redfern. *Australian Geographical Studies* 38, 3, 291–305

Shaw W.S. 2006. Sydney's SoHo Syndrome? Loft living in the urbane city. *Cultural Geographies* 13, 182–206

Short J.R. 1996. *The Urban Order: An Introduction to Cities, Culture and Power.* Oxford: Blackwell

Sibley D. 1995. *Geographies of Exclusion: Society and Difference in the West.* London: Routledge

Simmel G. 1995 [1903]. The metropolis and mental life, in P. Kasinitz (ed.), *Metropolis: Centre and Symbol of Our Time.* New York: New York University Press

Sissons J. 2005. First Peoples, Indigenous cultures and their futures. London: Reaktion Books Ltd

Smith N. 1979. Toward a theory of gentrification: a back to the city movement by capital not people. *Journal of the American Planning Association* 45, 538–548

Smith N. 1982. Gentrification and uneven development. *Economic Geography* 58, 139–155

Smith N. 1984. *Uneven Development: Nature, Capital and the Production of Space*. Oxford: Basil Blackwell

Smith N. 1987. Of yuppies and housing: gentrification, social restructuring, and the urban dream. *Environment and Planning D: Society and Space* 5, 151–172

Smith N. 1996. *The New Urban Frontier: Gentrification and the Revanchist City*. London: Routledge

Smith N. and Williams P. (eds) 1986. *Gentrification of the City*. Australia: Allen and Unwin

Smith S. 1989. *The politics of 'race' and residence: citizenship, segregation, and white supremacy in Britain*. Cambridge: Polity Press

Smith S. 1990. Social geography: patriarchy, racism, nationalism, *Progress in Human Geography* 14, 261–271

Smyth R. 1998. From the empire's 'second greatest white city' to multicultural metropolis: the marketing of Sydney on film in the 20th century. *Historical Journal of Film, Radio and Television* 18, 2, 237–262

Soja E. 1990. Heterotopologies: a remembrance of other spaces in Citadel LA. *Strategies* 3, 6–39

Soja E. 1996. *Thirdspace: Journeys to Los Angeles and Other Real-and-Imagines Places*. Oxford: Blackwell

Soja E. 1999. In different spaces: the cultural turn in urban and regional political economy. *European Planning Studies* 7, 1, 65–75

Sorkin M. 1992. *Variations on a Theme Park: The New American City and the End of Public Space*. New York: Noonday Press

South Sydney City Council 1999. *Draft Plan of Management: The Wilson Brothers Site and Yellowmundee Reserve, Caroline Street, Redfern, Sydney*. Sydney: South Sydney City Council

Spivak G. 1985. Can the subaltern speak? Speculations on widow sacrifice. *Wedge* 7, 8, Winter/Spring, 120–130

Spivak G. 1987. *In Other Worlds: Essays in Cultural Politics*. New York: Methuen

Spivak G. 1990. *The Post-colonial Critic: Interviews, Strategies, Dialogues*, ed. Sarah Harasym. New York: Routeldge

Stephenson P. 2003. New cultural scripts: exploring the dialogue between indigenous and 'Asian' Australians. *Journal of Australian Studies* 77, 57–73

Stratton J. 1990. *Writing Sites: A Genealogy of the Postmodern World*. Ann Arbor: University of Michigan Press

Sui D.Z. 1999. Postmodern urbanism disrobed: or why postmodern urbanism is a dead end for urban geography. *Urban Geography* 20, 4, 403–411

Taylor K. 1994. Things we want to keep: discovering Australia's cultural heritage, in D. Hedon, J. Hooton and D. Horne (eds), *The Abundant Culture: Meaning and Significance in Everyday Australia*. Australia: Allen and Unwin

Thomas N. 1994. *Colonialism's Culture: Anthropology, Travel and Government*. Cambridge: Polity

Thompson B. and Tyagi S. (eds) 1996. *Names We Call Home: Autobiography on Racial Identity*. New York: Routledge

Thompson D. (ed.) 1995. *The Concise Oxford Dictionary of Current English*. 9th edn, Clarendon Press, Oxford.

Thrift N. and Walling D. 2000. Geography in the United Kingdom 1996–2000. *Geographical Journal* 166, 2, 96–124

Tunbridge J. and Ashworth G. 1996. *Dissonant Heritage*. London: Wiley

University of Sydney 1985. *Report on the Development of the University Site. 1961–1985*. Sydney: Office of the Deputy Vice-Chancellor, University of Sydney

Urry J. 1990. *The Tourist Gaze*. London: Sage

Vipond J., Castle K. and Cardew R. 1998. Revival in Inner Areas. *Australian Planner* 35, 4, 215–222

Wasserman R. 1984. Re-inventing the New World: Cooper and Alencar. *Comparative Literature* 36, 2, Spring, 130–145

Wasserman R. 1994. *Exotic nations: Literature and Cultural Identity in the United states and Brazil, 1830– 1930*. Ithaca, New York: Cornell University Press

Weaver R.C. 1948. *The Negro Ghetto*. New York: Harcourt Brace

Wiegman R. 1999. Whiteness studies and the paradox of particularity. *Boundary 2*, 26, 3, 115–150

Wilton R. 2001. Colouring special needs: locating whiteness in NIMBY conflicts. *Social and Cultural Geography* 3, 303–21

Winders J. 2003. White in all the wrong places: white rural poverty in the postbellum US South. *Cultural Geographies* 10, 45–63

Wirth L. 1995 [1938]. Urbanism as a way of life, in P. Kasinitz (ed.), *Metropolis: Centre and Symbol of Our Times*. New York: New York University Press

Wray, M. and Newitz, A. 1997. *White Trash*. New York: Routledge

Wright P. 1985. *On Living in an Old Country: The National Past in Contemporary Britain*. London: Verso

Wright P. 1997. The ghosting of the inner city, in L. McDowell (ed.), *Undoing Place?: A Geographical Reader*. London and New York: Arnold

Young I.M. 1990. *Justice and the Politics of Difference*. Princeton, New Jersey: Princeton University Press

Zukin S. 1982. *Loft Living: Culture and Capital in Urban Change*. Baltimore: Johns Hopkins University Press

Zukin S. 1986. Gentrification: culture and capital in the urban core. *Annual Review of Sociology* 13, 129–137

Zukin S. 1995. *The Cultures of Cities*. Cambridge, Massachusetts: Blackwell

Index